Abstract

In November 2011, a utility-scale solar array became operational in the most unlikely of places, at Brookhaven National Laboratory in Eastern Long Island, New York. The Long Island Solar Farm project came together as a joint effort of five very different interest groups: a federal agency, a research institution, an electric utility, a private business, and the general public. The project is remarkable for three major reasons: first, it is the largest utility-scale solar power plant in the Eastern United States; second, it is a commercial project built on federally administered public lands; and third, the project was very unlikely to have started in the first place.

The process by which the Long Island Solar Farm was developed is intricate and unusual. This reflects many of the nuanced conditions that made siting the solar farm on federal property a unique opportunity for very different groups. Though many of these nuances make the Long Island Solar Farm difficult to imitate as a template, the research aspects of the project make it a trailblazing resource to inform future development of photovoltaic (PV) solar projects in the East. Furthermore, the innovation in attitude it took to develop this project serves as an excellent model for large-scale solar power development and public-private partnerships in general.

Figure A-1. Arial View of the Long Island Solar Farm at Brookhaven National Lab

Executive Summary

The Long Island Solar Farm (LISF) is a remarkable success story, whereby very different interest groups found a way to capitalize on unusual circumstances to develop a mutually beneficial source of renewable energy. The uniqueness of the circumstances that were necessary to develop the Long Island Solar Farm make it very difficult to replicate. The project is, however, an unparalleled resource for solar energy research, which will greatly inform large-scale PV solar development in the East. Lastly, the LISF is a superb model for the process by which the project developed and the innovation and leadership shown by the different players.

Unusual Circumstances: The Long Island Solar Farm exists at a nexus of very favorable conditions that were each crucially important to its development. The project began with what some might assess as an unnecessarily bold request for proposals from the electric utility, which was looking for a very large amount of PV solar power. That, in turn, opened the door for companies to consider developing a single-source utility-scale array on densely-populated Long Island, where the vast majority of existing PV solar comes from very small distributed arrays. Due to very limited land constraints on the island, BP Solar looked at the possibility of building a project on federal lands at Brookhaven National Laboratory (BNL). All parties involved were quick to recognize that the proposed site was replete with fortunate circumstances:

- No transmission was necessary; an existing substation was immediately next to the site
- Availability of large tracts of unused land on federal property in close proximity to load
- BNL is a "Superfund site," which produced an existing body of thorough environmental data
- BNL's history as a military base created a trove of historical and cultural information
- The federal facility's mission was such that a commercial project could be made mutually beneficial.

An Unlikely Project: The Long Island Solar Farm was, by all accounts, a tremendous leap outside the comfort zone for each of the participating groups: a federal agency (U.S. Department of Energy), a research institution (Brookhaven National Laboratory), an electric utility (Long Island Power Authority), a private business (BP Solar), and the general public. Each of these groups could have easily denied the project or refused to participate on the grounds that such a project was not what their organization "usually does" or it would "set a wrong precedent." Had any single group decided not to participate, or wholly object to it—reflecting their most likely attitude—then this project would have never taken place. Instead, and despite the odds, each of these very different interest groups found a way to make the project succeed.

Innovation in Attitude: Each of the groups exhibited excellent leadership within their respective organizations. No single dominant leader emerged in the consortium, which fostered a truly collaborative effort. Each group faced significant risks in this endeavor, but found innovative ways to safeguard their interests and mitigate those risks. The hallmark agreements of the project are: the use of an easement versus a lease for the land, the use of consideration as a means of compensation, the development of a dedicated research array, and the creation of a robust natural resources benefits package to mitigate environmental impacts.

A Trailblazing Resource: The conditional nuances of the Long Island Solar Farm make the development of a similar PV solar power plant very unlikely. This project does, however, serve as a very important data center, whereby enormous amounts of information about the array are being collected for analysis and research. The information gleaned from the solar farm and the dedicated research array will spur innovations in meteorological forecasting, PV solar technology, component systems design, large-volume storage, and utility-scale grid integration, among other areas of research. Though it may be difficult to build another Long Island Solar Farm, the lessons learned from its exposure to the dynamic conditions on the Atlantic coast certainly make it an invaluable resource to help inform future development of PV solar power in the Eastern United States.

A Model for Development: The organizational schematic in Figure ES-1 shows the dynamic results of the collaborative effort. Unsurprisingly, there were a great deal of negotiations behind each of these interactions and agreements. Each group had a sense for the project's potential, though, which kept them committed to its success. The outcome was the successful accomplishment of a very ambitious project that has already generated significant benefits for each of the parties involved. It is this process by which the Long Island Solar Farm was conceived, negotiated, and developed that make it an excellent example of public-private partnerships in general.

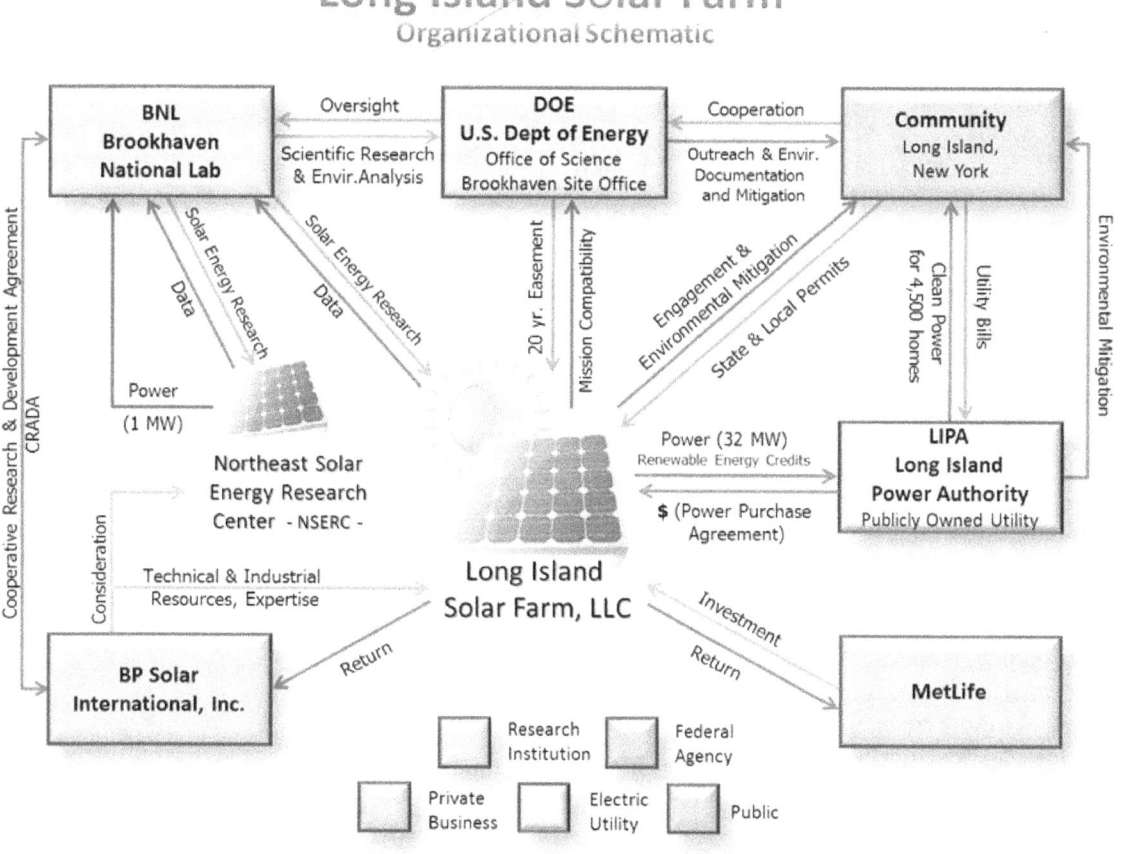

Figure E S-1. Organization of the Long Island Solar Farm Participants

iii

Table of Contents

List of Figures

List of Tables

Introduction

Generating power by converting sunlight into electricity is not a new concept. Neither is generating solar power at the utility scale. Photovoltaic (PV) panels have been collecting photons for decades, and PV arrays have been generating power for utilities since the first megawatt-scale solar farm was built in Sacramento, California, in 1984 (Green Energy News 2009).

What is new, however, is the accelerating demand for clean energy, particularly PV solar energy. The modern surge for solar is, in part, driven by rising demand for electricity and increasing environmental costs associated with conventional fuels. In recent years, large-scale solar energy development has also been invigorated by the economic forces of technological innovation, falling costs of production, and political support in the way of renewable energy standards and goals.

As a result, numerous large-scale solar projects have taken root domestically and internationally, and they are continuing to grow. Understandably, many of the earliest projects were developed where the sun shines the most. In the United States, for example, that has been primarily in the western portion of the country, more specifically, in the Southwest. The Western United States is certainly attractive for solar development for many reasons: land tends to be cheap, environmental impacts tend to be less complex, the population is comparatively less dense, the solar irradiance is high, humidity is low, and the weather is predictably cloudless for most of the year.

Though the conditions out West are rather ideal, large-scale solar power is still very much a viable source of renewable energy in a myriad of conditions and locations. In the Eastern United States, particularly in the Northeast, for instance, the conditions are dramatically different from the Southwest—land tends to be expensive, environmental impacts tend to be very complex, the population is more dense, the solar irradiance is comparatively less, humidity can soar, and the weather is highly variable and extremely difficult to predict. Even though the conditions may not be as ideal as those found in the West, economic forces are spurring the feasibility of PV solar power development in the East.

The Eastern United States has thus become a new frontier for utility-scale PV-solar energy development. Accordingly, there are few projects substantial enough to demonstrate PV solar energy's viability as a large-scale renewable resource for utilities in the East.

The Long Island Solar Farm (LISF) has strong potential to serve as that example. The LISF is a mere 60 miles from downtown New York City and is the largest solar array east of the Mississippi River. At the time of its development, it was ranked 5th among the largest PV power plants in the United States (and tied for 25th in the world) in terms of capacity (Lenardic 2012).

Though there are many lessons learned in the conception and development of the Long Island Solar Farm, it is best viewed as an example of the bold innovation in attitude it took to create a mutually beneficial source of clean, renewable power in what might otherwise seem an unlikely place for a large-scale PV solar farm.

Project Participants

- Federal Agency   U.S. Department of Energy

- Research Institution   Brookhaven Science Associates, BNL

- Private Industry   BP Solar, Inc.

- Electric Utility   Long Island Power Authority

- Environment   The Long Island Pine Barrens Society

U.S. Department of Energy, Office of Science, Brookhaven Site Office

> The mission of the U.S. Department of Energy (DOE) is to ensure America's security and prosperity by addressing its energy, environmental, and nuclear challenges through transformative science and technology solutions. The department's strategic goals to achieve the mission are designed to deliver results along four strategic themes: Transformation of the Nation's Energy System; The Science and Engineering Enterprise; National Security; and Management and Operational Excellence.

The mission of the Office of Science is to foster, formulate, and support forefront basic and applied research programs that advance the science and technology foundations necessary to accomplish DOE missions. The Office of Science is responsible for six multi-program national labs across the country and is the single largest supporter of basic research in the physical sciences in the United States.

The DOE Brookhaven Site Office is located at Brookhaven National Laboratory in Upton, New York. The mission of the Brookhaven Site Office is to manage and administer DOE's performance-based contract with the Brookhaven Science Associates for the management and operation of Brookhaven National Laboratory.

The Brookhaven Site Office supports the Office of Science mission to deliver the remarkable discoveries and scientific tools that transform our understanding of energy and matter, and advance the national, economic, and energy security of the United States (Office of Science website, accessed August 2012).

Henceforth in this document, "DOE" will be used to refer to the Department of Energy's Office of Science, Brookhaven Site Office at Brookhaven National Laboratory.

DOE's principals in this project are:

- Robert J. Gordon, Director of Business Management
- Louis Sadler, Chief Counsel

Brookhaven Science Associates, Brookhaven National Laboratory

Brookhaven Science Associates (BSA) was established for the sole purpose of managing and operating Brookhaven National Laboratory.

Formed as a 50/50 partnership between Battelle and the Research Foundation of State University of New York (SUNY) on behalf of Stony Brook University (SBU), BSA is the legal entity responsible for leading BNL successfully through the 21st century.

Both SBU and Battelle are strongly motivated to ensure Brookhaven's success. Being Brookhaven's closest university neighbor, SBU is the single largest user of BNL facilities; BNL facilities and scientific staff are essential to the vitality of the university's intellectual life and to the impact of many of its research programs.

BSA is governed by a 16-member board of directors: five appointed by the Research Foundation, five by Battelle, and one each from six of the nation's premier research universities (Columbia, Yale, Cornell, Harvard, Princeton, and the Massachusetts Institute of Technology).

Henceforth in this document, "BNL" or the "lab" will refer to the lab site itself or the institution. "BSA" refers to the Brookhaven Science Associates staff, whose principals in this project were:

- J. Patrick Looney, Ph.D., assistant laboratory director for policy and strategic planning, Sustainable Energy Technologies Department

- Robert J. Lofaro, group leader, Renewable Energy Group, Sustainable Energy Technologies Department

- Timothy Green, Ph.D., manager, Environmental Protection Division

BP Solar

British Petroleum Solar International, Inc., commonly referred to as "BP Solar," is a former subsidiary of British Petroleum. It was a manufacturer and installer of photovoltaic solar cells headquartered in Madrid, Spain, with production facilities in India and the People's Republic of China. Operating since 1981, the company became wholly owned by BP in the mid-1980s and was closed on Dec. 21, 2011 (Bergin 2011).

BP Solar was selected from a competitive process after Long Island Power Authority (LIPA) published a request for proposals for PV solar development in May 2008. As of this publication, BP is still under contract with DOE and BSA regarding the Long Island Solar Farm and its complementary research array. BP Solar is working to sell their easement rights and transfer obligations to a third party.

BP Solar's principal in this project was Richard Chandler, Project Manager

Long Island Power Authority (LIPA)

In May 1998, the LIPA became Long Island's primary electric service provider. LIPA is a non-profit municipal electric provider, owns the retail electric transmission and distribution system on Long Island, and provides electric service to more than 1.1 million customers in Nassau and Suffolk Counties and the Rockaway Peninsula in Queens. LIPA is the 2nd largest municipal electric utility in the nation in terms of electric revenue, 3rd largest in terms of customers served, and the 7th largest in terms of electricity delivered. In 2010, LIPA outperformed all other overhead electric utilities in New York state for reduced frequency of service interruptions and ranked 2nd for shortest duration of service interruptions. LIPA does not provide natural gas service or own any on-island generating assets (LIPA website, August 2012).

LIPA's principal in this project was Michael J. Deering, Vice President of Environmental Affairs.

The Long Island Pine Barrens Society (LIPBS)

The Long Island Pine Barrens Society is an environmental education and advocacy organization focusing on protecting drinking water and preserving open space, especially in Long Island's Pine Barrens.

> The LIPBS is a nonpartisan, not-for-profit organization dedicated to the study, appreciation, and protection of these unique woodlands. Founded in 1977, the society has become one of Long Island's most effective champions of preserving natural resources through sound land use. The society influences the public debates regarding land use, conservation, and other environmental issues through its scientific research and programs of public education and advocacy.

The LIPBS should not be confused with the Central Pine Barrens Joint Planning & Policy Commission, commonly referred to as the "Pine Barrens Commission." The Commission was established by the New York State Long Island Pine Barrens Protection Act of 1993. It has jurisdictional legal authority over the land use in the Central Pine Barrens region. The Pine Barrens Society, by contrast, has no jurisdictional authority.

The LIPBS was the largest voice of opposition to the development of the Long Island Solar Farm. Its principal is Richard Amper, Executive Director.

The Central Pine Barrens Joint Planning and Policy Commission

In 1993, New York State's Long Island Pine Barrens Protection Act defined this region at the junction of the towns of Brookhaven, Riverhead, and Southampton. The 1993 act created a five-member Central Pine Barrens Joint Planning & Policy Commission and an advisory committee, and mandated the production and implementation of the Central Pine Barrens Comprehensive Land Use Plan, adopted in June 1995. Under New York Environmental Conservation Law Article 57, the commission is responsible to produce, implement, and administer a Comprehensive Land Use Plan. The act and the plan charge the commission with the combined duties of a state agency, a planning board, and a park commission:

- Land use review, permitting, and enforcement authority in the Central Pine Barrens, along with the local municipalities

- Establishment and operation of a transferable development rights and conservation easement program

- Coordination of public lands stewardship and management on a regional basis.

The Long Island Solar Farm (LISF) at a Glance

The LISF is a 32-megawatt (MW) solar photovoltaic power plant built through a collaboration that included BP Solar, LIPA, and DOE. The LISF, located on the Brookhaven National Laboratory site, began delivering power to the LIPA grid in November 2011 and is currently the largest solar photovoltaic power plant in the Eastern United States. It is generating enough renewable energy to power approximately 4,500 homes, and is helping New York state meet its clean energy and carbon reduction goals (BNL, August, 2012).

Project Location	• Brookhaven National Laboratory, Upton, New York
Land Area	• Approximately 200 acres
Property Type	• Federal land (administered by the DOE)
Host	• DOE issued a 20-year easement to LISF, LLC to use land at Brookhaven National Laboratory
Developer/Owner	• LISF, LLC is jointly owned by BP Solar and MetLife, Inc.
Workforce	• 200+ FTE during construction, two during operation
Operator	• The day-to-day operation and maintenance of the LISF are contracted to True South, Inc.
Technical Design	• 164,312 ground-mounted panels (24 panels / rack) • 180° azimuth • Fixed-tilt at 27° (for optimized capacity during summer months) • 25 power blocks with dual inverters
Special Design Feature– Dedicated Research Array	• The 32-MW LISF was built in response to a request for proposals (RFP) issued by LIPA. Part of the agreement for use of the federal property included the development of a dedicated research array, called the Northeast Solar Energy Research Center (NSERC). The LISF generates 32 MW of commercial power. The 1 MW generated by the NSERC is retained and used on site at Brookhaven National Laboratory.
Technology	• Polycrystalline silicon solar PV modules
Commissioned	• Nov. 3, 2011
Lifespan	• Estimated to be 40 years (2051)
Peak Capacity	• 32 MW (AC)
Annual Energy Output	• 44 gigawatt (GW)/hours (estimated annual average), equivalent to the annual usage of approximately 4,500 homes
Utility	• LIPA purchases 100% of the LISF output, according to a 20-year power purchase agreement it has with LISF, LLC (expires November 2031)
Power Distribution	• To LIPA residential and commercial customers in Eastern Long Island

Map

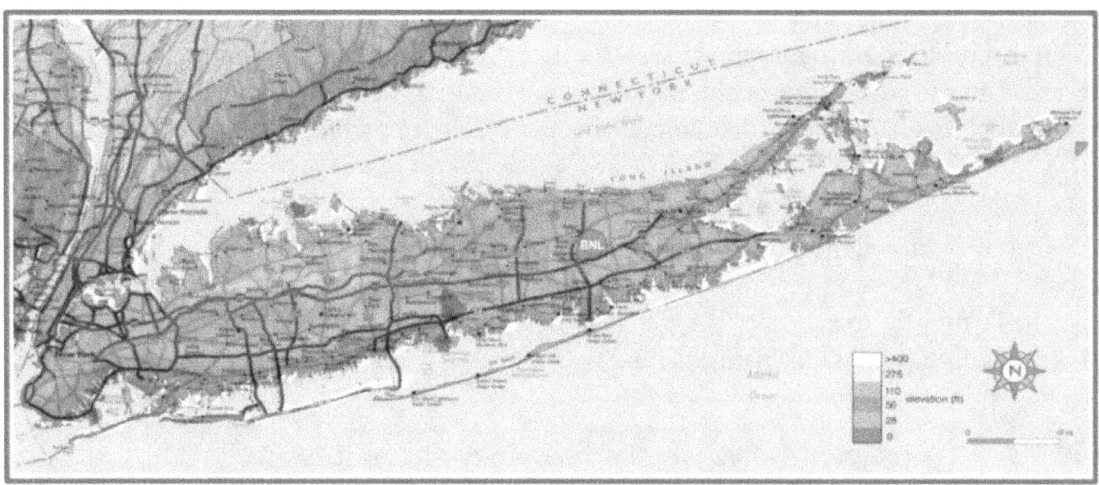

Figure 1. Location of Brookhaven National Laboratory

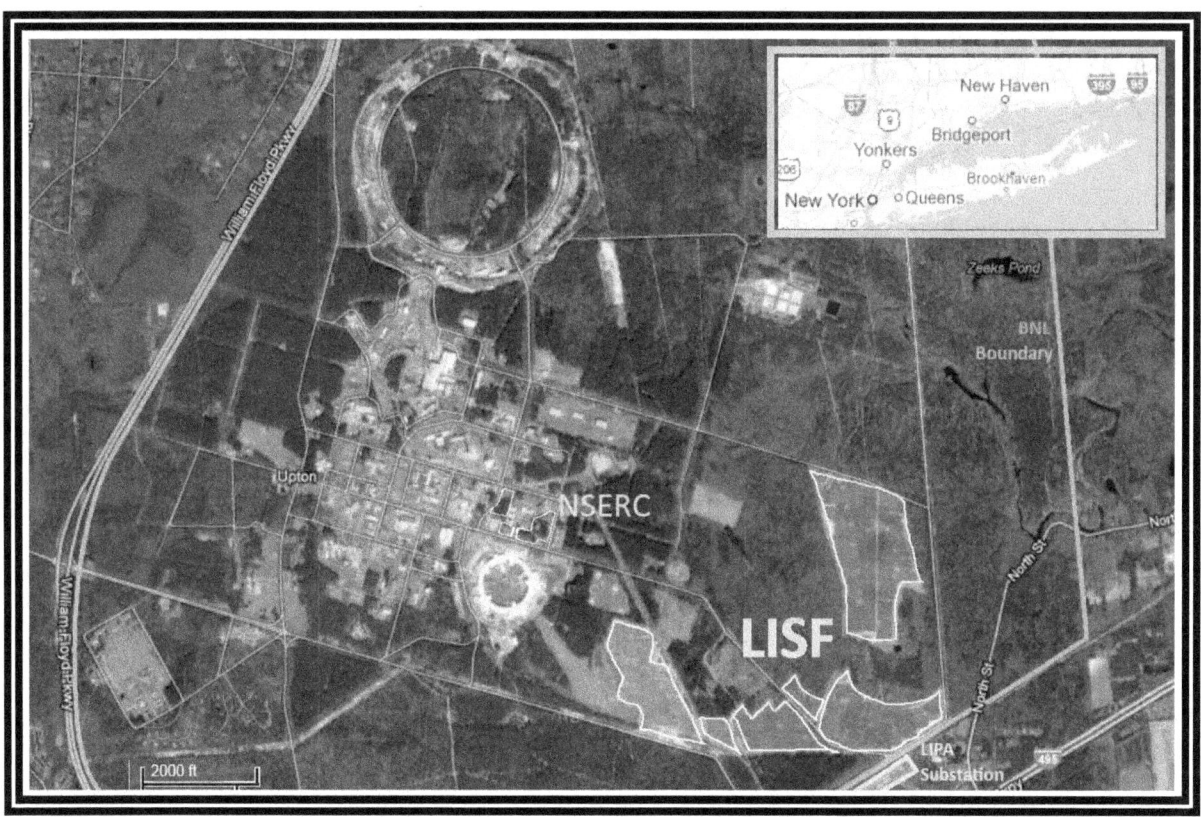

Figure 2. Satellite View of Brookhaven National Laboratory

Historical and Contemporary Context

In order to understand the development of the Long Island Solar Farm and to appreciate its size and scale, it is important to consider the historical and contemporary context of the project. This section provides a broad overview of the Long Island Solar Farm as it relates to PV solar development in the rest of the world and across the United States. It also establishes a backdrop of the state and local conditions that helped shape the development of the project.

Throughout the World

The use of photovoltaic power generation has been accelerating in the United States for the last decade. This is a trend that mirrors the development of PV systems around the globe, which has also been rapidly expanding. As a share of the world's installed PV capacity, the United States is ranked 4th (tied with Spain), behind Germany, Italy, and Japan (EPIA, 2012).

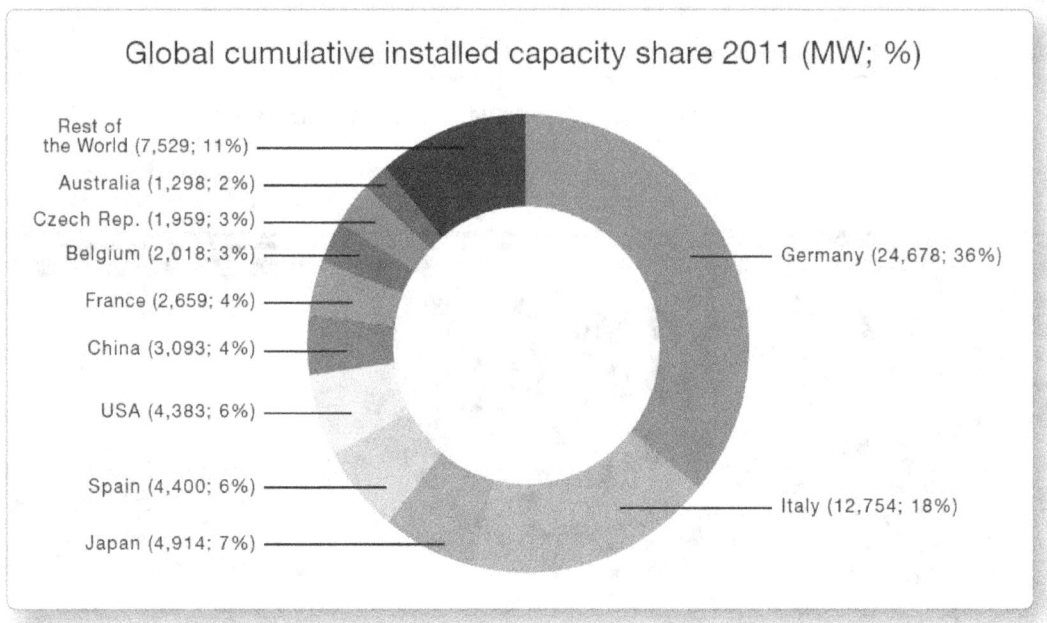

Figure 3. Global cumulative installed capacity share 2011

Naturally, the size of the individual PV systems varies dramatically throughout these different parts of the world. Countries with lesser overall installed capacities tend to rely on small-scale distributed generation. More ambitious countries, on the other hand, have developed very large utility-scale arrays in conjunction with their distributed programs. In this context on the world stage, the Long Island Solar Farm is tied for 25th in terms of power output.[1] Considering the thousands of generation facilities that have been built around the globe, the LISF is a remarkably large array.

---

[1] This statistic was taken at the time of its commissioning in November 2011. As of July 2012, the LISF is tied for 28th in the world due to larger projects being constructed in India, Europe, and the Western United States (Lenardic 2012).

On the domestic front, the Long Island Solar Farm establishes itself among the largest PV solar arrays in the country. At the time of its commissioning, the LISF was the 5th largest PV plant in the United States and by far the largest facility in the East.[2]

PV Solar Power Generation Facilities in the United States

Rank	Power	Country	Location	Description	Commissioned
1	48 MW	USA	Boulder City, NV	Copper Mountain Solar Facility	2010
2	45 MW	USA	Kettleman Hills, CA	Avenal Solar Facility	2011
3	42 MW	USA	Sonoran desert, AZ	Mesquite Solar I	2011
4	35 MW	USA	Mosca, CO	San Luis Valley Solar Ranch	2011
5	32 MW	USA	Long Island, NY	Long Island Solar Farm	2011
6	30 MW	USA	Colfax Co, NM	Cimarron Solar Facility	2010
7	30 MW	USA	Webberville, TX	Webberville Solar Farm	2011
8	25 MW	USA	Arcadia, FL	DeSoto Next Gen Solar Energy Center	2009

Table 1. PV Solar Power Generation Facilities in the United States

One of the most notable distinctions in Table 1 is that the Long Island Solar Farm is the only PV facility of its size in the Northeast. It comes as no surprise that the largest arrays are concentrated in the Western states. The solar resource map of the United States clearly identifies the Southwest as the region with the most intense and consistent irradiance, which produces the best solar energy yield. This, among other factors, makes large-scale PV projects very attractive in the West.

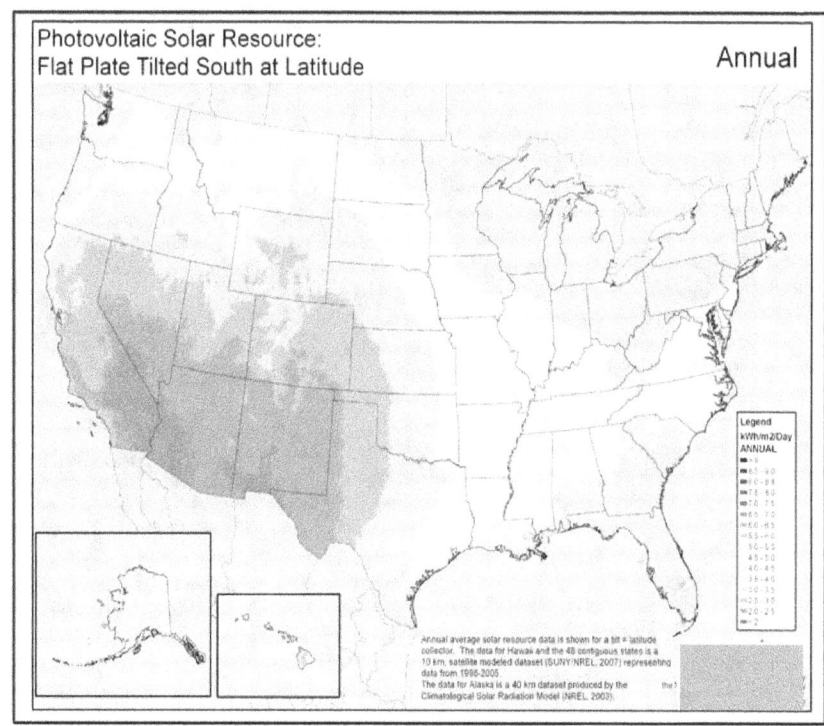

Figure 4. Photovoltaic solar resource: flat plate tilted south at latitude

---

[2] This statistic is from the time of its commissioning in November 2011. As of July 2012, the LISF is ranked the 7th largest in the country due to very large projects that have come online in Arizona and Nevada (Lenardic 2012).

9

Policy and Law

There are, of course, other forces driving utility-scale solar development. A myriad of political influences, federal and state policies, and technological reasons are making large-scale solar more attractive in places where the conditions are less than ideal. For starters, the Energy Policy Act (EPACT) of 2005 was a pivotal piece of federal energy legislation—the first comprehensive energy law in 13 years. EPACT authorizes and encourages, through the use of incentives and grants, energy production, conservation, and improved efficiency in permitting new facilities. It also helps set conditions to spur investments in renewable energy development through the use of tax credits and other incentives (LIPA 2004). Federal tax credits, like the production tax credit (PTC), helped catalyze investment and development in a broad sense. A provision of the PTC expired at the end of 2011, which caused a rush to get projects, like the Long Island Solar Farm, online before that deadline for credit application expired.

Federal Agency Initiatives

Federal agencies have taken strides to unlock the potential of large-scale solar opportunities around the country. The U.S. Department of the Interior, for example, has developed solar development zones (SDZ) on public lands in the West. SDZs are selected areas that minimize the impacts of development. The environmental characteristics of SDZs have already been (or are being) inventoried so as to facilitate shorter permitting timelines.

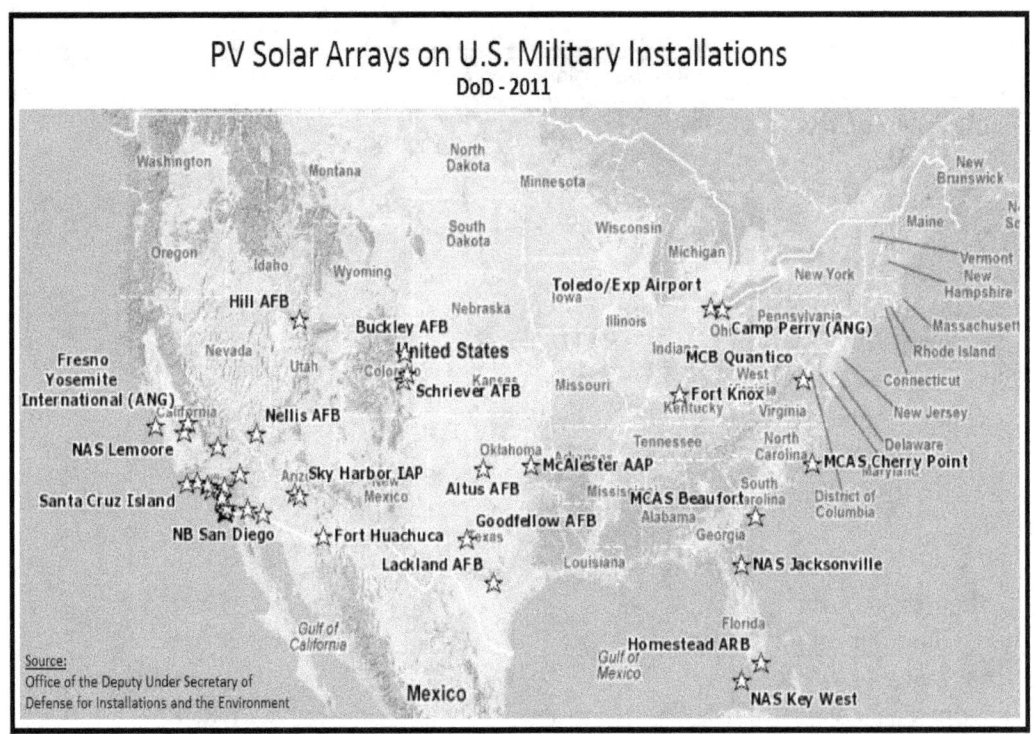

Figure 5. PV solar arrays on U.S. military installations

Other federal initiatives have been made by the Department of Defense. The DOD is the country's single largest consumer of energy and spends $4 billion every year on facility energy consumption (SERDP Website). In 2002, Congress appropriated funds for the DOD to do a comprehensive energy assessment, which it completed in 2005 (Snook 2007). Subsequently, through programs like the Strategic Environmental Research and Development Program

(SERDP) and the Environmental Security Technology Certification Program (ESTCP), the DOD is investing in innovative technologies, including PV solar, to help it achieve its renewable energy goals.[3] The programs use federal installations as test beds for cutting edge technologies developed by researchers from industry, universities, and federal agencies (SERDP 2012).

In concert with the DOD, DOE is looking at its facilities with an eye for renewable energy resourcing. In 2009, DOE created an Energy Parks Initiative (EPI) to take a focused look at using existing landfills, brownfields, greenfields, and federal facilities as opportunities for renewable energy development. EPI is designed to foster effective partnering of DOE, other federal agencies, private industry, state and local governments, local and regional communities, and stakeholders (Gilbertson 2009).

In the last decade at the national level, actions by Congress and initiatives taken by major federal agencies, such as the U.S. Department of Interior, DOD, and DOE, have significantly helped drive innovation and expansion of PV solar development throughout the country.

Among the States

At the state level, many legislatures have adopted or are considering the use of renewable portfolio standards (RPS). These RPSs are designed to steer electricity resource development away from fossil energy toward cleaner sources of power.

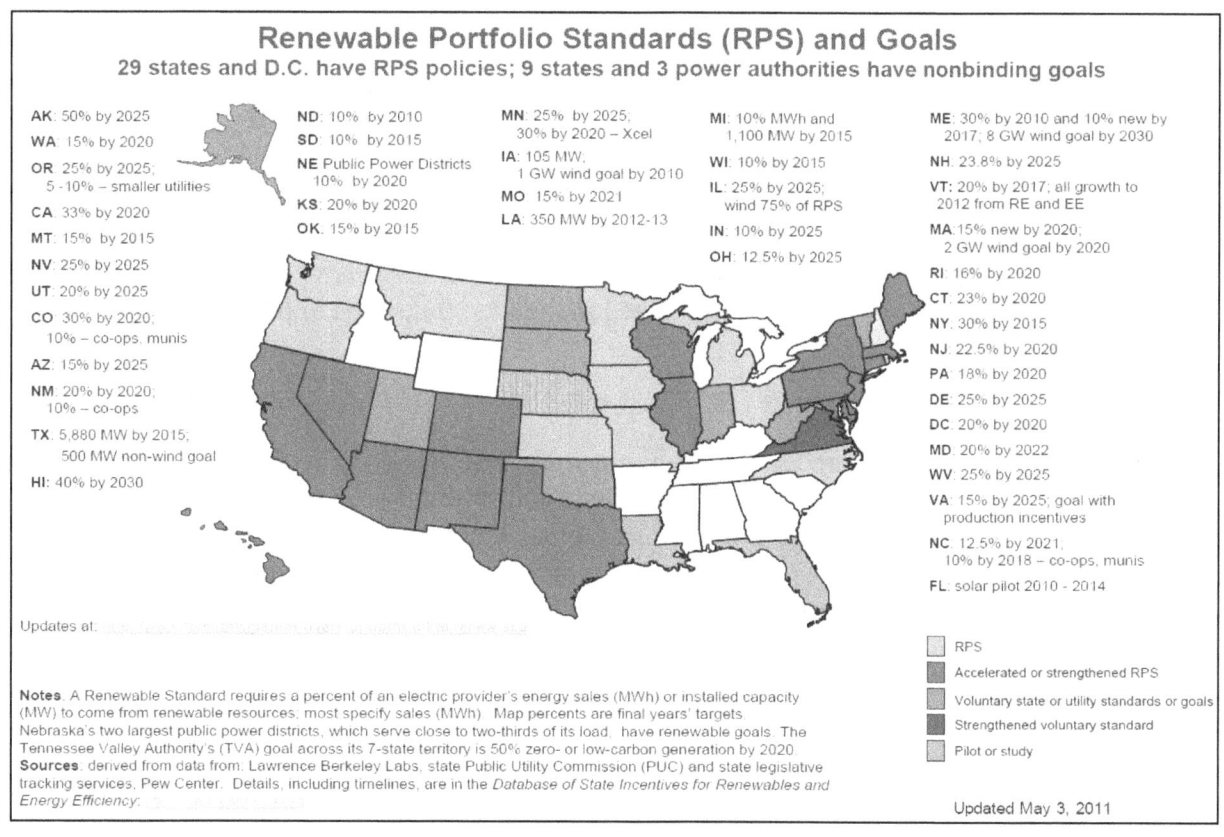

Figure 6. Renewable Portfolio Standards (RPS) and Goals

---

[3] The DOD is currently at approximately 10% toward a long-range goal of 25% by 2025 (SERDP 2012).

Figure 6 (FERC 2011) shows an uneven distribution of RPSs across the country, with varying degrees of intensity, both in terms of percentage goals and time horizons. Those states that have adopted RPSs, or similar programs, are generally those that have large and increasing load projections, and a comparative scarcity of power resources. The Southeast, in contrast, has thinner populations and access to very cheap power. As a result, those states are less inclined to expand renewable resources to higher proportions of their energy mix.

In New York

New York state—and the Northeast generally—is subject to very heavy and increasing loads associated with dense populations and urban development. The region also relies heavily on conventional fuel resources, which are volatile and costly. In confronting these conditions, New York has emerged as one of the most aggressive states in the country with regard to the breadth and boldness of its policies toward renewable energy development. For example, New York's RPS—to generate 30% of its electricity from renewable resources by 2015[4]—establishes the highest percentage of renewable power deployment in the country, matched only by California and Colorado (FERC 2011). New York has also established clean energy initiatives, green power purchasing programs, and various other polices to encourage residential and commercial development of PV solar (and other renewables). Governor Cuomo's New York Sun Initiative puts increased funding toward expanding solar power in the state. New York is definitely facing the sun and striving to be a leader in PV solar energy development.

On Long Island

LIPA, the state-owned utility for all of Long Island, has been a leader for solar development. In 2000, LIPA began a forward-leaning program of incentives for residential and business-scale solar projects with capacities between 50 kilowatt (kW) and 3 MW. LIPA was very bold in issuing an RFP in 2008 for projects that could deliver 50 MW of solar PV. This was the first of its kind in New York state. LIPA has also developed a program for solar feed-in tariffs, to standardize integration to the grid. This is another first-of-its kind program in New York (Deering 2012). LIPA is also consistently ranked in the top 10 utilities in the country in total PV solar capacity (SEPA 2012).

Local Environmental Issues

There is a very strong awareness of environmental issues on Long Island. The very fact that it is an island amplifies the notion that its environmental resources are well-defined. The natural environment is not boundless, as it may seem on the mainland. Instead, its environmental resources are very finite as they extend into the sea in all directions. In addition to this, Long Island rests atop a sole-source aquifer, which means that nearly all of the drinking water on Long Island comes from wells in the ground. The people of Long Island, in general, have a heightened concern about pollution of their waterways and protection of their limited natural resources.

---

[4] Starting at 19% in 2004 (DSIRE 2012).

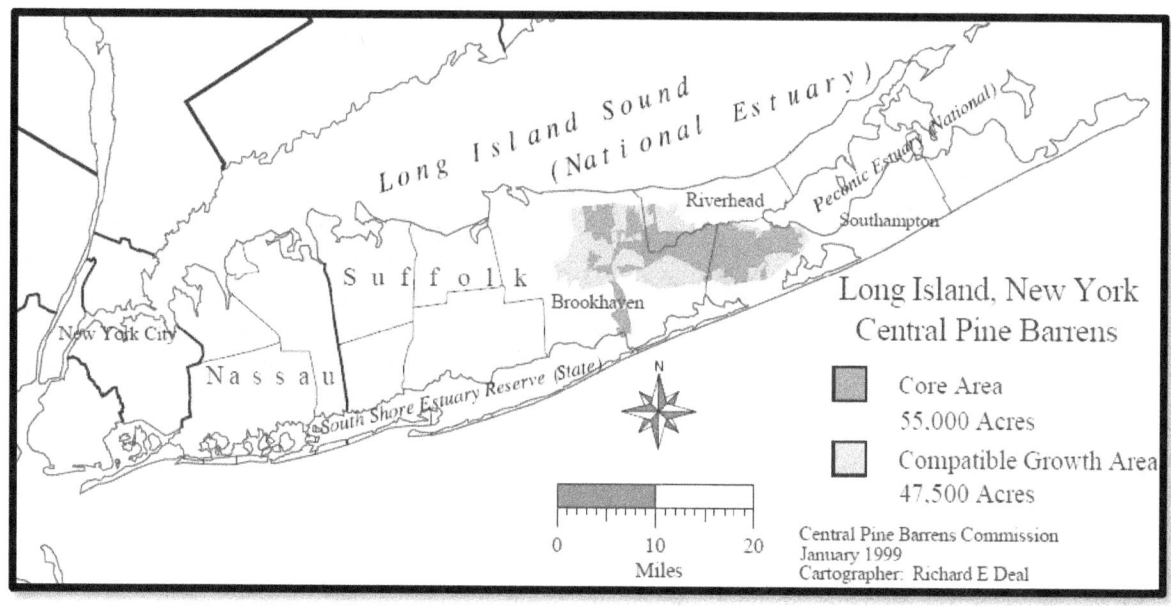

Figure 7. Map of Long Island, New York, Central Pine Barrens

Evidence of this is that the combined funding levels—from town, county, and state laws—for preservation and conservation of natural resources in Suffolk County (the eastern two-thirds of Long Island) is $2.2 billion since 1977, which is more than that of 45 other states (Amper 2012).

Another seminal environmental achievement for Long Island was the legal protection of the Central Pine Barrens, the island's largest natural area and last remaining wilderness. The New York State Long Island Pine Barrens Protection Act of 1993 identified core preservation areas and established the Pine Barrens Commission with jurisdictional authority to manage compatible growth areas (PBC, 2012). The law has been very successful and serves an important feature of Long Island's acute environmental sensitivity.

Summary

At first glance, it may seem peculiar that one of the largest PV solar arrays in the United States—and in the world—is located on Long Island. However, in descending from a worldwide view to a local perspective, it is clear that the concept of a utility-scale PV solar array on Long Island was not an accidental or haphazard coincidence. Rather, there was a nexus of key factors that were essential to its creation: federal agencies looking for innovative development opportunities, states pushing to incentivize solar energy, and communities wanting clean sources of electric power. These conditions created fertile ground for a project like this to take root.

Project Concept

How it all Got Started

For nearly a decade, LIPA had established a suite of PV solar initiatives geared toward small-scale projects, primarily at the residential level. In wanting to move solar to larger-sized commercial projects, in 2008 LIPA issued an RFP for 50 MW of PV solar power from anyone who could supply it. This was a very bold request, and the first of its kind in New York state.

According to Michael Deering, LIPA's vice president for environmental affairs, LIPA was agnostic about whether that 50 MW could be generated by one single array or if it could be done with several smaller arrays distributed around the island. The solar RFP was an open question, and they waited to see what the industry's responses would be. Nevertheless, given the resource constraints on Long Island and the large volume of power they were requesting, LIPA generally expected that they would get those 50 MW in a distributed manner. The RFP opened in April 2008 and closed that following August. LIPA received a total of 35 proposals. Unsurprisingly, most of them were for distributed development of the 50 MW.

Seizing the Initiative

Naturally, one of the biggest resource constraints for PV solar is the availability of land. Generating a lot of PV solar power requires a lot of acreage. On a densely populated island with very finite space, there were few tracts of land that could host a very large array in a cost-effective way. Recognizing this, LIPA released another RFP to identify any land owners on Long Island who would be interested in hosting a PV array. LIPA made the list of potential land owners accessible to the respondents of the solar RFP, but not vice versa (Deering 2012). Again, the majority of the land owners could only support small portions of the 50 MW projects.

Soon after the solar RFP was publicized, both LIPA and BP Solar recognized that one of the single largest land-owners was Brookhaven National Laboratory. In separate calls that happened to be around the same time in May 2008, both LIPA and BP Solar contacted the lab. LIPA made a call to Dr. Looney, of BSA, and BP Solar called Bob Gordon, director of the Business Management Division for DOE at Brookhaven National Laboratory. In both cases, they posed the question, "Would Brookhaven National Laboratory be willing to host a solar farm?" (Looney, Deering, Gordon, Chandler 2012).

Using Federal Lands

Discussions began immediately at the lab between BSA and DOE, who analyzed how such a project might fit the mission of the DOE and serve the interests of the lab. In June, LIPA published on its website a formal letter from DOE. The letter proposed the possibility of using federal property at Brookhaven National Laboratory to companies participating in the solar RFP. This was a very attractive prospect for many companies who were looking for sufficient acreage for a large PV solar project.

What was not attractive to many companies, however, were the unusual stipulations that DOE put on its proposal. In order to use the site, a project would have to agree to two key provisions:

- The project must include a research component, in order to fit with DOE's mission for the lab.

- The land-use agreement would be an easement, not a lease or a license.

These conditions repelled nearly all the would-be developers. The research component of the deal was unusual, to be sure. This was a venture for private business to develop solar power for commercial use. Most companies quickly rejected the superfluous costs that the research component presented.

DOE's insistence on using an easement was widely regarded as very unattractive to investors.[5] In general, developers of projects like this aggressively seek out licenses or leases for full control rights of the land during the time of the agreement. This gives them rock-solid rights to use the property, which provide a critical foundation of certainty for their business models.

DOE could not support such a wholesale transfer of control that was associated with a lease or license. It was essential that DOE retained access to the land under the array. In a practical sense, DOE would need to retain access rights in order to manage the research aspect of the project. More importantly, though, in a long-term strategic sense, DOE needed to reserve ownership of the land in case of a change in federal priorities for the site (Easement Outgrant 2010). To the vast majority of developers, such an easement represented an intolerable degree of uncertainty for their long-term business models. The only developer that did not shy away from those stipulations was BP Solar.

LIPA's Choice—Selection Criteria

After thoroughly reviewing the details of each of the 35 proposals, LIPA made its selection on which companies would fulfill its request for 50 MW. The review process was very rigorous in quantitative and qualitative analysis. The selection criteria was based on a number of factors, including the quality of the developer, cost-benefit analysis, location of the site(s), likelihood of obtaining the requisite permits, interconnection logistics, etc. (Deering 2012).

One of the key aspects of each proposal was the ability of the project developer to demonstrate site control; that is to say that the developer had (or could obtain) total control over the proposed site. Just like the developers, LIPA wanted to minimize or nullify those risks associated with land use in order to safeguard its power purchase agreement. Because BP Solar was the only company willing to take the business risks associated with DOE's stipulations, it was the only proposal that incorporated land use at Brookhaven National Laboratory (Deering 2012).

It was important that there was only one company that was willing to work with DOE. On the one hand, it was easy for DOE to make a commitment to BP Solar, at least in terms of entering negotiations with them and nobody else. On the other hand, that relationship demonstrated to LIPA the requisite level of site control for the proposal. In other words, if multiple companies

---

[5] DOE informed its use of an easement, in part, by a similar land-use arrangement made for a PV solar developer on federal property administered by DOE at the National Renewable Energy Laboratory in Colorado (Sadler 2012).

had submitted proposals that claimed they could use land on Brookhaven National Laboratory, it would have demonstrated that none of them really had any sort of commitment from DOE. This in turn, would have raised the level of uncertainty for LIPA to an intolerable level, and they would have rejected those proposals for a utility-scale array (Deering 2012).

In consideration of all these factors, LIPA ultimately selected two companies, BP Solar and enXco. BP Solar was awarded a contract to construct a single array that would produce 32 MW.[6] The other 18 MW would come from enXco, who would develop a series of distributed arrays built over large parking lots in several places around Long Island.

Power Purchase Agreement

Upon selection, BP Solar established the "Long Island Solar Farm, LLC" as the legal representation of the solar farm as an entity unto itself. LIPA negotiated the terms of a power purchase agreement (PPA) with LISF, LLC. to provide the energy produced and renewable energy credits (REC) from the solar farm. The costs to LIPA under the PPA for the energy produced by LISF, LLC. total approximately $298 million, which include the costs of interconnection. The time horizon of the PPA is 20 years from the date of in-service, which was Nov. 3, 2011.

The results of the PPA amount to about $0.60 per month for the typical residential customer. According to LIPA's Chief Operating Officer, Michael Hervey, "This agreement not only allows us to provide clean energy to our customers, but it also delivers price stability for our customers in an energy market where oil and gas prices remain volatile" (LIPA 2011).

Shared Investment—LISF, LLC

BP Solar could have done this project on its own but chose to bring in a partner as an investor. Once the PPA was signed and the project design was about 30% complete, the project was stable enough to bring an investor on board. MetLife became that partner and served as a passive hands-off investor with no other stake in the project other than its financial investment. Even though MetLife owns the majority stake of the Long Island Solar Farm, they are still an insurance company. They do not have the engineering, procurement, or construction wherewithal to take an active role in the project. BP Solar, therefore, serves as the leader and primary resource of consideration, according to the terms of the easement (Chandler 2012). In order to accommodate the dual ownership, the solar farm itself is established as a limited liability company, officially entitled the "Long Island Solar Farm, LLC."

---

[6] LIPA's original breakdown of the 50 MW had 37 MW coming from BP Solar and 13 MW from enXco. LIPA reallocated 5 MW from BP Solar to enXco during the design phase of both projects, leaving 32 MW for the LISF (Deering 2012).

Process

* Major Milestones

	Date	Action	By	For
2008	April 22, 2008	LIPA Issues a Solar RFP	LIPA	Anybody
	May, 2008	BP Solar approaches BNL with an interest to use BNL land; LIPA also approaches DOE with interest for land use at Brookhaven	BP Solar & LIPA	BNL & DOE
	June 30, 2008	DOE/BHSO Submits Hosting Letter - Proposing Access to BNL	DOE/BHSO	RFP Applicants
	August, 2008	Proposals Due to LIPA	Industry	LIPA
	November, 2008	LIPA Selects BP Solar's Proposal - PPA Signed	LIPA	BP Solar
2009	April, 2009	Presentation to Community Advisory Council	DOE - BHSO	Community
	June, 2009	Real Estate Appraisal Published	MacCrate Associates LLC	BP Solar
	September, 2009	Solar Easement Draft Published	DOE	BP Solar
		Appraisal Value Negotiation Meeting	BHSO, BNL & BP Solar	
	December, 2009	NEPA: Environmental Assessment Complete	DOE	BP Solar
		FONSI	DOE	LISF
2010	January 20, 2010	Easement Granted	DOE	LISF
	February 3, 2010	MOU Signed	DOE & BP Solar	LISF
	October, 2010	Easement Management Plan Signed	DOE / BHSO	LISF
		Construction Activities Begin	LISF	DOE/BHSO
	December, 2010	Begin Development of CRADA 1 (Cooperative Research and Development Agreement) for Instrumentation of the LISF	BNL & BP Solar	LISF
2011	August, 2011	Liability Insurance Issued		
		Building Code Compliance Approval	Sigma Energy Engineering	BP Solar
	September, 2011	Begin Developmnet of CRADA 2 (Cooperative Research & Development Agreement) for the Dedicated Research Array	BNL & BP Solar	LISF
		NEPA Review of Research Array Site Complete	DOE	BP Solar
	November 3, 2011	Flip the Switch: Full Commercial Operations Begin at the LISF	BP Solar	LIPA
	December, 2011	Easement Management Plan Signed	DOE / BHSO	BP Solar
		CRADA 1 and CRADA 2 are executed	BNL & DOE	LISF
		Press Release: BP Solar Will Go Out of Business	BP Solar	BP
2012	Spring 2012	Continue to Develop CRADAs 1 and 2, Transition Ownership of the LISF	BP Solar	BNL
	November, 2012	Begin Construction of Research Array	BP Solar	BNL
2013	March, 2013	Flip the Switch: Research Array is Operational	BP Solar	BNL
	November 3, 2031	Easement Expires	DOE/BHSO	LISF

Figure 8. Generalized Timeline of Development for the Long Island Solar Farm

No Easy Easement

The first and most important aspect of negotiations for this project was the development of the easement agreement between BP Solar and DOE. As stated before, the easement was the best of three options for DOE, and the worst of the three for BP Solar. Lou Sadler, DOE's legal counsel, explained the basic pros and cons of each from the DOE's perspective.

Lease—The basic terms of the lease included things intolerable to DOE:

- DOE would lose control of the site for the time period of the agreement
- Any consideration would have gone straight to the U.S. Treasury
- No opportunity for scientific research
- Would greatly extend the timeline (involving formal public comment periods).

License—The basic terms of a license were similar in many ways to those of a lease:

- Both leases and/or licenses were attractive to investors because they are nice, neat, and locked in for long periods of time. This reduces risk in the analysis of the financiers.

Easement—This was the only way DOE was willing to move forward with the project:

- Easements are not typically used for projects like this (leases or licenses are predominantly used by investors).
- Easements generally allow for someone to have very limited use of a parcel of land; a common example is for the use of telephone poles and/or electric power lines to cross a particular area.
- The rationale was that, at its core, there wasn't much difference between the LISF project and the common application of an easement for development of transmission lines; the solar power array would be constructed over land that was otherwise unused.
- DOE also insisted on the use of an easement because they would retain:
  - Full ownership and access to the land
  - The compatibility of the project with DOE's mission, through the availability of in-kind consideration to be applied directly to the research agenda
  - The ability to meet the condensed timeline of the project, as set by LIPA.

The easement negotiations were somewhat tricky. BP Solar very much wanted the agreement to sound as much like a lease as possible, of course, to minimize their risk. In contrast, DOE insisted that it must retain access to and control of the land, and could not risk relinquishing it in a lease or license.

Consequently, as the two sides negotiated to common ground, they tried to narrow the gap of risk between a lease and an easement. Richard Chandler, project manager for BP Solar, described the negotiations:

Ultimately, it's not the label at the top of the document, but the rights and obligations outlined in the document that are most critical. So long as we were okay with the responsibilities and tenets of the agreement, we were willing to move forward. The feeling, of course, was mutual on the other side of the table. But, we wanted to narrow the language of the agreement to ensure, as much as possible, that if BP Solar dedicates its resources, then the land would not be taken away (Chandler 2012).

Mr. Chandler alludes to the key feature of the negotiations—trust. In other words, in order to make a final deal, that gap of risk between a lease and an easement needed to be spanned with mutual trust that couldn't be captured per se on a legal document.

On one side of that gap, BP Solar needed to trust that DOE would not pull the land out from under the array for any unexpected reason in the next 20 years at a minimum (the term of BP Solar's PPA with LIPA). In practicality, the solar farm site represents a small portion of all available (or unused) lands at the lab. Thus, it would be extremely unlikely that DOE would have a reason to have to go through the massive undertaking of dismantling the solar farm and either repurposing that specific land for other development or restoring it to its natural state. Such a scenario was so unlikely that BP Solar was willing to trust DOE's assurances that they'd let the Solar Farm exist on that site for the duration of the PPA. This is a strong measure of good faith, as the easement explicitly authorizes DOE to do what it wants with the land beneath the solar site at any time (Easement Outgrant 2010).

On the other side of the gap, DOE had to trust that BP Solar would uphold its end of the deal and provide in-kind consideration to develop the dedicated research array. BP Solar wanted—and was afforded—specific language in the agreement that enabled them to liquidate the value of consideration and write a check to the U.S. Treasury (Gordon, Lofaro 2012 and Easement Outgrant 2010). This was not in the best interest of DOE or the lab. If BP Solar did that, it would completely undermine the research component of the solar farm, and disassociate the entire project from the mission of DOE and the lab. Nevertheless, and despite the significant risk, DOE trusted that BP Solar would follow through with its commitment to provide consideration instead of cash.

In the end, the keystone of the whole easement—and consequently the entire solar farm project—is that level of trust over crucial interests at the core of the agreement.

Reinforcing Commitments, Insuring Trust

As a compromise, written explicitly in the terms of the easement, BP Solar insisted on retaining the ability to write a check to the U.S. Treasury for the balance between the estimated total value of consideration and the actual in-kind consideration they could deliver. To illustrate, if BP Solar was committed to a package of in-kind consideration (of tools, equipment, special expertise, etc.) valued at $100,000 but only contributed $70,000 worth of necessary resources, then BP Solar would be able to write a check to the U.S. Treasury for the remaining balance of $30,000.

This posed a significant liability for DOE and the lab because, at any time, BP Solar could choose to liquidate its commitment of consideration and pay it to the U.S. Treasury, which does not directly benefit the interests of DOE and BNL (Lofaro 2012). It would undermine the research component of the project, which was the hallmark feature and the essential viability of the project from DOE's perspective.

In order to ensure that BP Solar followed through, DOE insisted on an up-front commitment of funds from BP Solar that would be dedicated expressly to the development of the dedicated research array. This was a commitment that took some leveraging of BP Solar, who didn't think it was necessary for DOE to insist on such a down payment. Nevertheless, DOE wanted a tangible commitment to protect the development of the research array if BP Solar was inclined to back out of the project for some unforeseen reason.

Despite their verbal assurances that they would follow through in providing consideration, BP Solar ended up having to back out of their long-term commitment. The company declared it was going out of business soon after the LISF was built but before the design began on the dedicated research array. Were it not for this down payment to BSA, BP Solar would have been able to write a check to the U.S. Treasury for the difference. This would have technically fulfilled their legal obligations in the easement and enabled BP Solar to walk away (Sadler 2012). Of course, that disastrous scenario would have left DOE and BNL without a dedicated array for solar energy research.

BSA currently has that down payment of $2.3 million in holding (Gordon 2012). That investment is, at the time of this publication, driving BP Solar to fulfill its commitment to developing the research array before they completely decouple themselves from this project.

Project Design

The Use of Consideration

As stated previously, in order for a commercial project to be built on federal property at a research institution, it must incorporate a research component that would effectively support DOE's mission for the lab. Richard Chandler, project manager for BP Solar, recognized this early on. He saw it as an opportunity that could benefit not just the lab but also his own company. He also recognized that the federal agencies weren't as interested in monetary compensation as the landowners typically were in other solar projects. Under the terms of the easement, BP Solar worked very closely with BSA and DOE to develop the definitions of "consideration," which would serve as the exchange of in-kind contributions to a Cooperative Research and Development Agreement (CRADA). Consideration refers to compensation given by one or more parties to an agreement that can be in the form of goods, services, and/or money.

According to the easement, BP Solar is obligated to deliver consideration to DOE in return for the easement to use the land. The total valuation of the consideration in dollar terms is approximately $6 million over the 20-year easement. This has been estimated and generally agreed upon as the aggregate value of all equipment, labor, construction expenses, maintenance issues, land use, and costs associated with the research agenda (Gordon 2012). So, instead of BP Solar paying a full rent check to DOE for use of the land, BP Solar would provide other resources in the way of materials and expertise. Any balance left over would be paid by BP Solar directly to the U.S. Treasury. DOE, of course, prefers that the consideration be applied on site at the lab (Easement MOU 2010).

As Richard Chandler describes, "The use of consideration was a way to find common ground and create a mutually beneficial exchange of goods, services, and resources. This was a natural fit for both BP Solar and Brookhaven. Both BNL and BP Solar were very interested in Solar (research and development). The bottom line is that developers need to approach a project like this, dealing with multiple entities, particularly with federal agencies, with an appreciation for what the other side is looking for" (Chandler 2012).

Two-Pronged Approach

The quintessential research component of the Long Island Solar Farm was formally codified in the CRADA. This agreement was broken into two parts, CRADA 1 and CRADA 2, each with a different research focus:[7]

- CRADA 1: The inclusion of data collection instrumentation on the LISF

- CRADA 2: Development of a dedicated research array (NSERC)

---

[7] See the "Research Opportunities" section for more details.

Technical Design

Overall Description

The system is a large-scale commercial solar photovoltaic array of approximately 37 MW (direct current), and covers approximately 200 acres of the BNL federal site. Electricity generated by the system is connected into the regional utility power grid. The system is comprised of individual solar modules. Approximately 164,000 modules are used, and the current design has modules arranged into sub-arrays consisting of four vertical rows of six modules. The 24 modules are attached to a steel I-beam support structure that is anchored with two steel I-beam driven posts. These posts are driven approximately 15 feet (ft.) [4.6 meters (m)] below the surface of the soil. Strings of modules are wired together in series. Fused combiner boxes with disconnect switches connect strings of modules and lead to 32 concrete equipment pads of about 100 square feet each, placed around the array. Each equipment pad contains two inverters and one megavolt ampere (MVA) transformer. The wiring between the modules and equipment pads are trenched. Conductors combine at 15 kV outdoor, metal-clad, switchgear on a 250 square-foot concrete pad. A 13.8/69 kV transformer, with associated circuit breakers disconnect switches and a small (12"x24") control enclosure mounted on pads totaling 2,000 square feet, then steps up the power and supplies it via cable to the Long Island Power Authority 8ER Substation, located on the other side of the Long Island Rail Road (LIRR) track from BNL. The 69 kV lines, including three conduit lines, [one for fiber optics and two for power distribution (one active and one spare)], run under the Long Island Railroad tracks (Easement Outgrant, 2010).

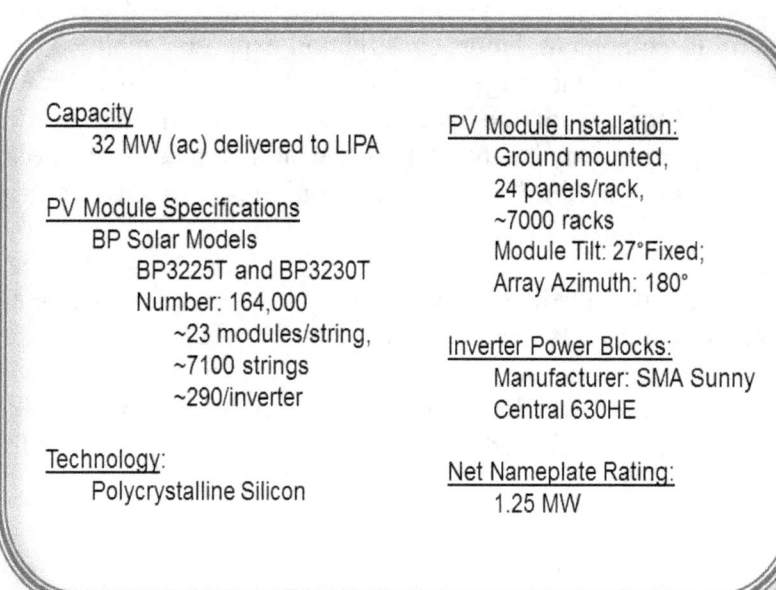

Figure 9. System Specifications

Capacity
    32 MW (ac) delivered to LIPA

PV Module Specifications
    BP Solar Models
        BP3225T and BP3230T
    Number: 164,000
        ~23 modules/string,
        ~7100 strings
        ~290/inverter

Technology:
    Polycrystalline Silicon

PV Module Installation:
    Ground mounted,
    24 panels/rack,
    ~7000 racks
    Module Tilt: 27°Fixed;
    Array Azimuth: 180°

Inverter Power Blocks:
    Manufacturer: SMA Sunny
    Central 630HE

Net Nameplate Rating:
    1.25 MW

Technology

The array is polycrystalline photovoltaic cells on a fixed array, oriented due south and tilted at 27 degrees. This technology and configuration optimizes the array for maximum production output at its particular geographic location and also during the summer months, when loads on Long Island are peaking. This was also the most efficient technical design in terms of cost-benefit analysis (Chandler 2012).

22

Patchwork Design

The easement authorized BP Solar to use oddly shaped portions of the property. At first glance, this arrangement appears to be very random. It was, however, the result of very deliberate environmental considerations and impact mitigations.[8]

This patchwork of six separate areas presented significant challenges for design. It required BP Solar to spend a lot of time reconfiguring the technical blueprint of the array, which increased the overall design cost. By comparison, the large-scale arrays built in the open deserts out West are typically laid out in nice rectangular blocks with lots of room to spare. On Long Island, however, it was a constant challenge to pack in as much power onto parcels of land that were both incongruous and very irregularly shaped (Chandler 2012).

Figure 10. Satellite View of the LISF and NSERC at Brookhaven National Laboratory

Transmission

It was very fortunate that LIPA already had an existing substation immediately adjacent to Brookhaven National Laboratory. As shown in Figure 10, even more convenient was that the substation was adjacent to the southeast corner of BNL property, which was otherwise unused. This made the southeast corner of BNL an ideal site for the array. These serendipitous conditions completely eliminated the need to build attendant transmission infrastructure for the project. For all intents and purposes, they could literally plug the whole thing right into the grid.

Though the distance for transmission was negligible in terms of its impact on the whole project, it was not zero. Approximately 900 ft. (274 m) of transmission cable were installed between the BNL property line and the LIPA substation 8ER point of connection to the grid. The cable, consisting of two conduit lines (one active and one spare), had to cross the LIRR tracks between the lab property and the substation. The stretch of transmission line that crossed the LIRR easement was constructed belowground using horizontal directional drilling (EA-1663 2009).

---

[8] See the section on "Environmental Impacts" for more details.

23

When asked whether BP Solar would have considered building such a large array if that substation was not there, Richard Chandler explained, "It would be difficult to answer. Undoubtedly, having to create transmission infrastructure would have added another layer of complexity to the project, but in the end, the answer to this question would come down to a standard cost-benefit analysis." The phrase, "another layer of complexity" may be an understatement, depending on how far a transmission connection would have to be. Given the location of the lab property, surrounded by state-protected wilderness, endangered species habitat, and very expensive private lands, it would have been another major project in its own right to design, site, permit, and build attendant transmission infrastructure. It certainly would have added substantial costs of time and money to the overall project (Chandler 2012).

Interconnection to the Grid

The LIPA substation required only very minor modifications to connect the LISF to the energy distribution grid. The substation modifications included the addition of two 69 kV disconnect switches, a 69 kV gas circuit breaker, 69 kV potential transformer, revenue metering, metering potential transformers and current transformers, and related control and protection relaying (Figure 11). All modifications occurred entirely within the substation footprint (EA-1663 2009).

There were no significant technical challenges associated with the interconnection with the grid. Once the design was 30% complete, LIPA, BP Solar, and New York Independent System Operator (NYISO) conducted an 18-month study of the solar farm's impact on the grid and found there to be no significant issues. This was the first solar project integrated into the NYISO, using their standard forms and procedures for integration (Chandler 2012).

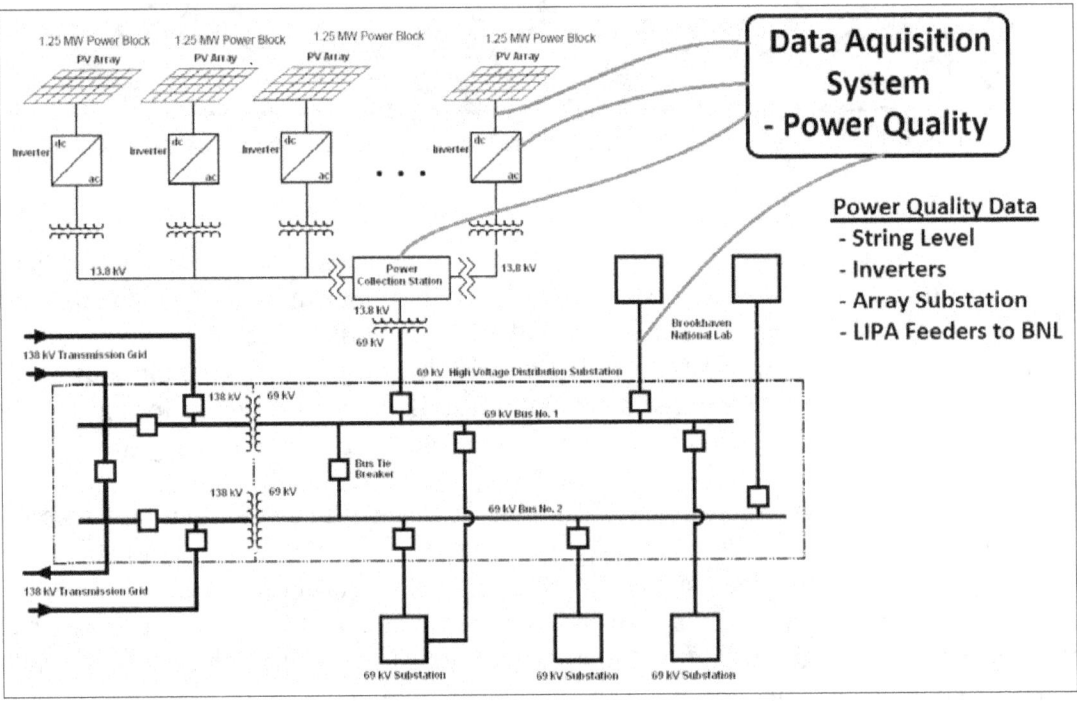

Figure 11. Grid Integration Diagram: Long Island Solar Farm

Vulnerabilities

Typically, solar power plants are most vulnerable to theft and vandalism. This is, in part, because they're typically in remote locations and require little-to-no operational supervision. Most stations are protected by fencing of some sort and many are monitored remotely with cameras (Chandler 2012).

The location of the Long Island Solar Farm insulates it from this type of vulnerability. The LISF is very difficult to access, in part because it is surrounded by the Pine Barren wilderness and also because it is located on the lab property, which is fenced-in, guarded, and patrolled.

As a consequence of its location, the Long Island Solar Farm is vulnerable to more extreme weather conditions than large arrays in other places. Long Island is prone to hail-producing thunderstorms in the summer, hurricane-force winds in the fall, and heavy snows in the winter.

In June 2012, for example, a portion of the array suffered a lightning strike during a particularly intense thunderstorm. Though the LISF has a 50-ft. lightning rod in the center of the array, the lightning hit one of the racks, busted dozens of fuses, and fried some electronics. This vulnerability to natural conditions represents a potential risk of high operational and maintenance costs. It also serves, however, as a uniquely valuable opportunity to research and model the effects of such varying conditions on a large array.

Public Engagement

Proactive From *(Before)* the Start

BP Solar put a lot of effort into making a good first impression with the community. Soon after LIPA selected their proposal, BP Solar organized an open house at a community center near the lab. The open house included all the proponents (LIPA, DOE, BP Solar, and BSA) and was open to everyone. The open house created a very good first impression, and the project did not receive any initial resistance.

Subsequent meetings and updates weren't standardized, but they were frequent enough to sustain the engagement between affected groups: DOE, BP Solar, LIPA, and the local community, elected officials, and environmental activist groups. BP Solar was very engaged from the onset. The company:

- Gave presentations to BNL Executive Committee and the Brookhaven Community Advisory Committee

- Visited all the local chambers of commerce

- Participated in BNL Summer Sundays (a series of open houses at the lab where the community is invited onto the lab for tours and educational programs); BP Solar's participation in this was not specifically about the solar farm project but more of a "PV 101" about solar energy to increase awareness of the technology

- Dispelled rumors and misinformation by posting billboards near/on site to explain to passersby details about the project

- Distributed flyers to everyone who was on or near the transportation route associated with the construction of the solar farm

- Solicited ideas from the public on how to improve the project and incorporated many of their suggestions into the project's development (Chandler 2012).

More broadly, although there was generally not a lot of media attention given to the project until it was near completion, a few reporters in the regional print media consistently published updates on the development of the solar farm. This was helpful to keep the public aware of the progress and status of the project. In general, the public engagement strategy was generally to lean forward and get out ahead of any issues. This active approach to engaging with the community was instrumental in defining the public message, communicating progress, and minimizing the grounds for backlash from political or special interest groups (Gordon 2012).

Richard Chandler concurs with his description of BP Solar's outlook on communication. He explains that the biggest takeaway regarding public relations was that, "It is really easy for misinformation to get out in front of a project, which could have negative and/or crippling effects on its development. We got out there early and often to get ahead of any of that to set the tone and control the message; to make sure the public was getting accurate, forthright information about the project's development." Both DOE and BP Solar recognized that this project could not go on without public support. Their proactive strategy of engagement ensured the public was a part of the development process, which significantly enhanced its chances of success.

Siting and Environmental Issues

Environmental History and Current Conditions

Superfund Site

All of BNL is classified as a Superfund site and is listed on the National Priorities List, which is administered by the Environmental Protection Agency. This classification makes BNL eligible for funding relating to environmental cleanup and mitigation of the effects of hazardous materials. This is primarily because of the types of pollutants that were discarded and the unique aspects of some of the decommissioned laboratory facilities on site. Brookhaven National Laboratory's Superfund classification has no direct impact on the easement, land use, or the LISF project. The only exception is that some parts of the BNL property are classified as off limits or unusable (Gordon 2012).

Citizens Campaign for the Environment (CCFE)

Adrienne Esposito, Founder and Executive Director

CCFE:

- Is a lobbying entity

- Has no legal authority

- Created a groundswell movement to support the project

- Is an advocate of LIPA's perspective

- Thought if this project did not happen now, then it would never happen—or a polluting alternative would eventually be built instead.

The Long Island Central Pine Barrens Society
(aka "The Pine Barrens Society" or "LIPBS")

Richard Amper, Executive Director

LIPBS:

- Is a lobbying entity

- Has no legal authority

- Felt the environmental benefits of this project were not worth the environmental costs of destroying Pine Barren lands

- Never endorses development but only actively opposes a development plan if it threatens to negatively affect water quality and Pine Barren ecosystems

- Has three basic levels of opposition:

  o Level 1—the society publishes a statement in opposition to a project.

  o Level 2—the society puts pressure on appropriate authorities to oppose a particular project.

  o Level 3 (litigation)—the society will file suit against a developer for violating the law. Projects very rarely come to litigation (approximately two cases per year) (Amper 2012).

DOE was not legally obligated to work with the LIPBS.

27

Central Pine Barrens Joint Planning and Policy Commission
(aka "The Pine Barrens Commission" or "PBC")

- Is a state entity with governing authority over the Pine Barren lands, according to the New York State Long Island Pine Barrens Protection Act of 1993

- Does not have jurisdiction over federal land on BNL property, though the Pine Barren habitat extends onto it.

Steven Jones was commission chairman at the time of development and strongly opposed the project as counter to the DOE's mission (Amper 2012).

Long Island Native Plant Initiative (LINPI)

- This group generally resisted the development of the LISF.

  They insisted on protecting the site from invasive species of plants and that the undergrowth of the LISF would be planted with native grasses only.

Divided Resistance

The environmental community was at odds over this project. On one hand, some people viewed the solar farm as a clean source of renewable energy that was better than an alternative source of power that would likely generate more pollution. This broad perspective of the project had its loudest voice in the Citizen's Campaign for the Environment (CCFE).

On the other hand, there were environmental activists who viewed this project as being destructive to the environment on which it would be built. This perspective's chief spokesman was Richard Amper, executive director of the Pine Barrens Society. He vehemently disagreed that renewable energy projects should come at such a high cost to the environment. He states, "The Society has worked with CCFE for twenty years and has agreed on just about everything, but this project posed the first dramatic rift between our organizations. Though we were always mutually very respectful, Adrienne [Esposito] and I had some challenging phone calls. We ended up agreeing to disagree" (Amper 2012).

Fundamental Environmental Debate:

> *"The Long Island Solar Farm destroys some aspects*
> *of the environment in order to preserve others."*
> ~ Richard Amper

The LISF indeed represents a cost of more than 160 acres of forest. Considering the cost, however, the carbon offset of the solar farm is at a much higher rate than the amount of carbon that area of forest could sequester. Plus, 100% of the land used in the project received some type of mitigation and an equal amount of lands were protected elsewhere, both on and off the property (Green 2012).

Richard Amper's perspective was that this project set up a false choice between "preserving natural ecosystems or building a solar energy farm." He states:

> This project extracted an environmental cost to build an "environmentally friendly" energy resource. The damage caused to the environment by the construction of the solar array is not offset by the solar farm's environmental benefits. The notion that we can develop renewable energy, but it has to be at a cost to the environment is a false premise, and that's what this project represents. LIPBS supports renewable energy and solar power, but not at the expense of Pine Barren ecological systems.

The alternative to the single-source utility-scale array would have been aggregating the necessary 150 acres of land from other areas, whether they be onsite at BNL or distributed throughout the rest of Long Island. LIPBS, therefore, much prefers the distributed model of solar development like the enXco project that paralleled the LISF development, according to LIPA's 2008 RFP (Amper 2012).

Legal Lines in the Sand

The environmental groups did not have any legal jurisdiction over the site. However, that didn't mean they were powerless to stop the project. Groups like the Pine Barrens Society could very well have killed the project by delaying it. They could have filed suit, requested an order of restraint, or drummed up popular resentment against the solar farm. Any of these contingencies, by themselves or in combination, could have significantly delayed the project. If delayed long enough, the project would have failed.

This project proved to have an interesting dynamic in the eyes of environmentalists. The stark contrast of perceptions for and against the solar farm created a tension in the environmental community that prevented a strong coalition of opposition from forming. This defeated any chance for environmentalists to create the political or social conditions that could suffocate the project.

Nevertheless, DOE was very forthright in dealing with the environmental concerns of the project. BP Solar and LIPA also took a forward-leaning approach to mitigating environmental impacts. In the end, the loudest sources of opposition (i.e., the Pine Barrens Society) were largely satisfied with the mitigation package from DOE, BP Solar, and LIPA.

Engaging the Opposition

Richard Chandler states, "By and large, most environmental groups were supportive of the project as a clean source of renewable power. The Pine Barrens Society, however, was the staunchest in its opposition. Our intent was not to oppose them but rather engage with them. We solicited their input and used their feedback to improve our project and mitigate its environmental impacts."

DOE established a similar disposition in dealing with environmental opposition. Bob Gordon explained that, "Right from the start we knew that the Pine Barrens Society would be the most vocal opponent to the plan. We met with them early and got them involved in the process right away, and made every effort to incorporate their ideas and suggestions."

This strategy was effective in engaging the project proponents with the LIPBS; however, Richard Amper disagreed. "BNL and DOE have always been very forthcoming with their plans. BNL (DOE) is the largest single owner of Pine Barrens habitat on Long Island. [The BNL-LIPBS] relationship had always been generally good up until the LISF project. The project had a cooling effect on that relationship" (Amper 2012).

Mitigation Package: Natural Resources Benefits

DOE, BSA, LIPA, and BP Solar were very interested in honoring the environmental considerations of opposition groups as much as possible. In working with the LIPBS and the Pine Barrens Commission, the project proponents put together a natural resources benefits package to mitigate the environmental impacts of the solar farm. The key tenets of this package include (CCFE, 2012):

- LIPA provides $2 million for open space preservation within the Central Pine Barrens region

- DOE preserves an additional 51 acres of property that builds upon the 500-plus acres it previously preserved in the 1990s

- BP Solar provides $75,000 for ecological habitat, research, and restoration.

$2 Million from LIPA

At first glance, this appears to be an ostensible "mitigation fee" put on LIPA. This is a misperception, according to Mike Deering. He explained that, "The $2 million was part of the regulatory process, as administered by the New York State Pine Barren Commission. They have jurisdiction over the development of the Pine Barren region, but not direct authority over the federal property. The commission suggested a 1:1 preservation match, and $2 million facilitated that. LIPA honored their suggestion and appropriated the money out of an existing fund for community benefits. That money was then dedicated for the purchase of lands for preservation."

Richard Amper agreed that was an important piece of mitigation. "Those funds essentially went through Brookhaven town and were subject to public input. The LIPBS contributed guidance on selection of lands for preservation. The final decision—to buy land in the Carmans River Watershed—reflected our input."

51 Acres from DOE

The Pine Barren Society had no legal influence over the project, but DOE worked to get the LIPB on board as much as possible. Richard Amper agrees by saying, "DOE, to their credit, were consistently and sincerely candid, accessible, and professional in their dealings with [the

LIPBS]. DOE took us on a tour of the facility and consulted with us for their site selection. DOE clearly wanted to minimize our unhappiness about the project. DOE was very upfront about their plans. Right from the start they set an inclusive tone." He also included a caveat, though. "Though they were inclusive, DOE made it clear very early on that this project was going to go forward, regardless of [LIPBS's] analysis that solar energy should be developed elsewhere and without damaging Pine Barren ecosystems."

DOE made concessions to the society throughout the process. One of these measures was to commit other portions of the BNL property to preservation. Thus, DOE classified 51 acres of wooded property as unavailable for future development by the lab.

It is important to note that this is not a binding agreement; DOE is in no way obligated to abide by that self-imposed measure. It has, nevertheless, incorporated that preservation area into its official long-term development plan.

$75,000 from BP Solar

This lump sum went to Long Island Native Plant Initiative (LINPI), which is a cooperative volunteer not-for-profit effort of more than 30 organizations, governmental agencies, nursery professionals, and citizens. It was founded in 2005 by Polly Weigand, technician for the Suffolk County Soil and Water Conservation District. LINPI continues to pioneer the collection and banking of seed from more than 35 native grasses, herbaceous plants, shrubs, and trees on Long Island. The grant from BP Solar was intended to go toward LINPI's ecological habitat, research, and restoration.

This was, more or less, a gift from BP Solar. According to Dr. Green, "The long-term benefit of this contribution, though, is that BNL will continue to have access to a source of native seeds that can be used in various projects on the Lab property (BNL would, of course, purchase that seed at market price). This will benefit vegetation programs at BNL, including on the land beneath the solar arrays."

Siting the Array in a Sensitive Environment

In addition to the marquis features of the natural resources benefits package, DOE and BP Solar put a lot of effort into making various other environmental considerations that shaped the final footprint of the solar farm.

The map in Figure 12 gives a broad perspective of some of the major environmental constraints on the lab property. The areas that were already open for development (shaded in pink) are very small and scattered and would not be able to efficiently support such a large system. The areas surrounded by the red borders indicate preferred or "core" Pine Barren ecosystems or compatible growth areas (CGA). In addition, the 1,000-ft. buffer established to protect the endangered tiger salamanders creates a random assortment of exclusion zones. This map reveals how the development of the solar farm resulted in a patchwork of six parcels of land (shaded in blue).

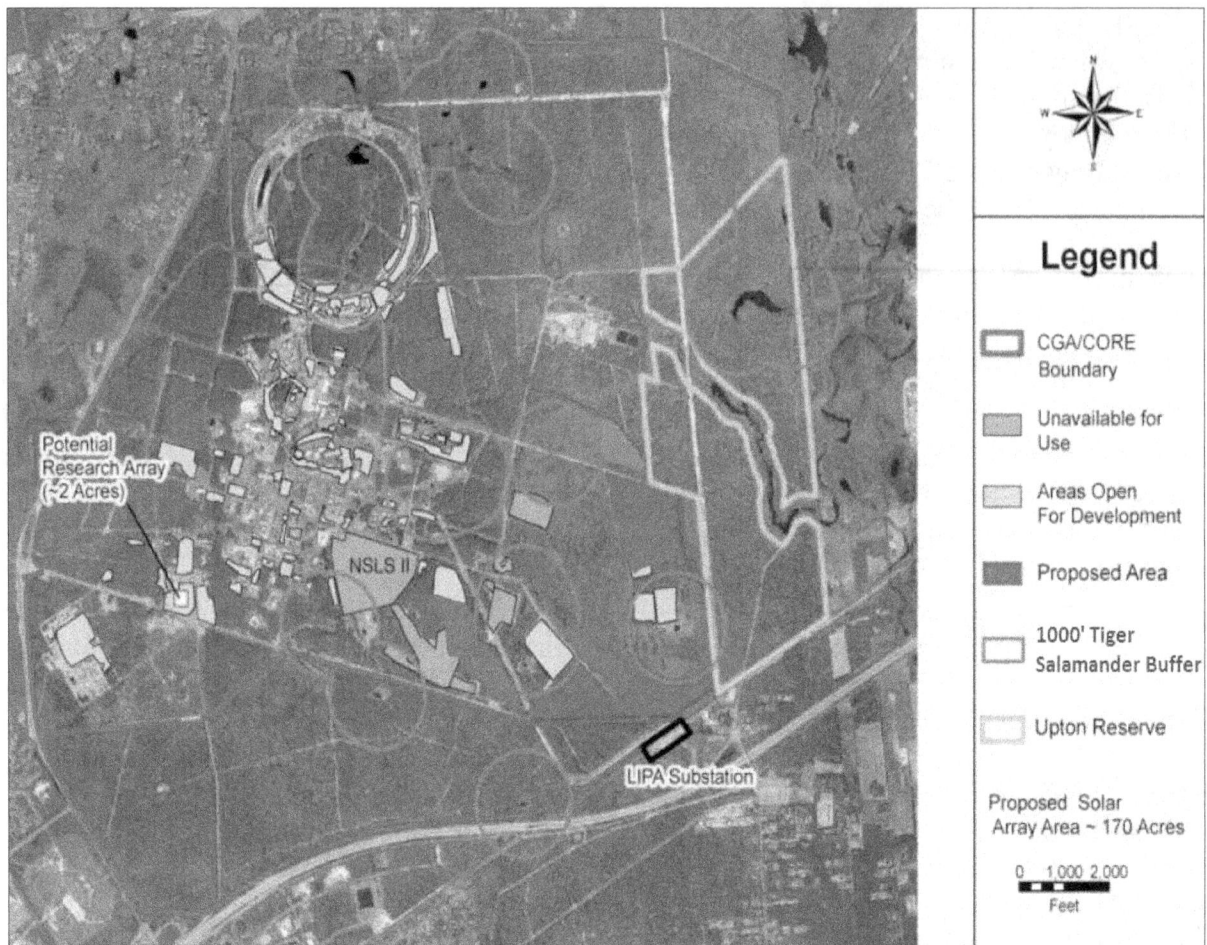

Figure 12. Proposed LIPA solar project

Figure 12 does not show all of the considerations that went into the site selection for the solar farm. It does, however, give a sense of the major constraints and the limits for development. The full suite of criteria demonstrates the complexity of the project and the high degree of concern that DOE, BSA, and BP Solar had in minimizing then environmental impacts where possible.

The following is a list of considerations made during site selection and design of the project:

- Considerations for selecting location:
  - o Avoided core preservation area of Pine Barrens
  - o Proximity to LIPA substation
  - o Avoids or reduces environmental and cultural resource impacts
  - o Utilizes already cleared or previously disturbed areas
  - o Limits impacts on BNL operations (away from utilities, traffic, and future science projects).

- Land type (total area is approximately 200 acres):
  - 35 acres of existing already-cleared lands
  - 5 acres of former tree nursery
  - 98 acres former World War I areas, with various levels of restoration
  - 62 acres that were already disturbed prior to World War I
- A portion of project area was moved to avoid 14 acres of higher-quality Pine Barrens habitat in CGA of the Pine Barrens
- The irregular layout of six separate parcels of land minimizes environmental impacts
- The project completely avoids development within core preservation area (CPA)
- Avoids wetlands and maximizes buffers around tiger salamander habitat
- Improves a small pond that is a breeding area for tiger salamanders
- Native grasses were planted to restore cleared lands
- Removes invasive plants and BNL actively prevents the establishment and spread of invasive species into the CPA
- No significant impact on groundwater
  - Total annual water use for maintenance less than 500,000 gallons
  - Native vegetation below the arrays will filter precipitation as it infiltrates ground.
- The project is not expected to impact surface water; current flow patterns will stay unchanged
  - Impervious surfaces increase by 10,890 square-feet (ft^2)
- Creates a deer-free area, enhancing habitat for other wildlife, and creating an ecological research opportunity
- Fencing is wildlife friendly, including gaps (approximately 5"x12") deliberately placed at the ground level, every 75 feet along the perimeter; this allows small animals and mammals unencumbered access in and out of the site
- Construction activities were timed to reduce disturbance to birds and wildlife
- Array connections to step-up transformers follow existing roadways and avoid wetlands.

Permitting

At first glance, the location of this project on federal property might appear to be a disadvantage. The project would be subject to multiple layers of review at the local, state, and federal levels, which could jeopardize the timeliness of the project's development. In the case of permitting, though, the unusual conditions of the property turned what could have been a disadvantage into a fortuitous benefit for the project.

Land Development at Brookhaven—A Storied History

Prior to becoming Brookhaven National Laboratory, the federal property was originally a military installation called Camp Upton, which opened in 1917. Shortly after World War II ended, Camp Upton was decommissioned, and the property was turned into a research institution under the jurisdiction of the Atomic Energy Commission, the precursor to the U.S. Department of Energy. From 1947 to 1989, the lab used and disposed of hazardous materials at the lab. The Environmental Protection Agency added the Brookhaven National Laboratory site to the Superfund National Priorities List on Nov. 21, 1989, because hazardous chemicals were found in the soil and ground water (EPA 2012).

> **BNL/DOE Land Preservation History**
> - 2,943 acres preserved (56% of current footprint)
> - 2,339 acres donated to NY State Parks – 1973
> - 74 acres miscellaneous land transfers
> - 530 acres dedicated as Upton Ecological & Research Reserve (10% of land area) – 2000
>
> **BNL Land Development over 60 year history**
> - 1,412 acres (27%) of current 5,265 acres
> - Original WW II area – 500 acres (10%)
> - Since 1947 – 912 acress (17%)
> - Long Island Solar Farm – 200 acres (<4%)
> - 30.6 % developed by 2011

This high-profile history of the lab site required several detailed reviews of the property through many different lenses. From a military perspective, the land has been closely surveyed for safety purposes, namely unexploded ordinance or hazardous material still buried in the ground. From an environmental perspective, the property has undergone a battery of very detailed reviews of its geological, hydrological, ecological, and biological characteristics. Even from a cultural perspective, the fact that it used to be a military installation has given cause for the State Historical Preservation Office (SHPO) to hold a record of surveys that catalogs military artifacts and important historical sites.

All these surveys and reviews over the years created a very thorough library of information with much of the critical information required by the permits. This turned out to be a great advantage for the development of the Long Island Solar Farm because all the pertinent information was readily available and comparatively little field work was necessary. According to Dr. Timothy Green, Environmental Protection Officer for BNL, "We studied the impacts and potential impacts of the project to the smallest degree. It was fortunate for the environmental review that there was already a large amount of detailed information about the site. This is largely because the lab used to be a military base, it is a Superfund site, and its environmental resources have been studied very carefully over the last 80 years."

Federal Permits

A full review was required under the National Environmental Protection Act (NEPA). This project required an environmental assessment, which is the second-most thorough level of review after an environmental impact statement (EIS). DOE was ultimately responsible for the NEPA certification. The formal environmental assessment was completed in December 2009. An official finding of no significant impact (FONSI) was published by the DOE shortly thereafter.

New York State Permits

- **State Environmental Quality Review Act (SEQRA)**—Because LIPA is a state-owned entity, this project was subject to the State Environmental Quality Review Act and needed a SEQRA certification. BNL conducted its own parallel review according to SEQRA, which required coordination between the lab and the utility. Both came to consensus on what was being analyzed and their respective results (Green 2012).

- **New York Department of Environmental Conservation (NYDEC)**—BP Solar worked very closely with the NYDEC regarding appropriate permits to build in proximity to particular species (in this case, key species include the box turtle and tiger salamander).

- **State Historical Preservation Office (SHPO)**—The historical nature of the site, particularly its military heritage, required a permit from the State Historical Preservation Office. The SHPO had to certify it had conducted an archaeological survey of the site.

- **Wild Scenic and Recreational Rivers Act**—This was a combined permit for freshwater wetlands and scenic or "viewshed" impacts. First, BP Solar agreed to do minor restoration of a freshwater pond in the center of the site. The pond was a breeding area for the tiger salamander, which is designated by New York State as endangered (NYSDEC 2012). Working in this area required a freshwater wetlands permit. Secondly, Brookhaven National Lab is very close to the Peconic River, the longest river on Long Island. Because of the potential effect of the solar farm on the wetlands and the viewshed associated with the river, this project needed a scenic rivers permit (Green 2012).

- **Radiological Contamination Site Interagency Agreement**—This was a multi-agency process to get NYDEC, EPA, DOE, and the Suffolk County Government to come to agreement on the status of environmental conditions and their effects on the development and operation of the LISF. Examples of some of the provisions include:

  o The standards to which the site had to be cleaned

  o Prohibitions on removing soils from the site

  o Protections for workers, etc.

- Though it was a challenge to coordinate among local, state, and federal agencies, the process was remarkably efficient. This was, in part, because most of the detailed information about the site was available at the onset, and the group could get right to work sorting out the details of how to manage those conditions. DOE took the lead to shepherd this group through discussions, and the committee reached agreement in less than nine months (Green 2012).

Local Permits

- Suffolk County requires a heavy equipment use and construction permit for projects like this within the county. Legally, though, BP Solar was not obligated to have that permit from the county government as the project was on federal land. Nevertheless, BP Solar obtained that permit from the county as a tangible way to demonstrate in good faith their responsibility as a "good neighbor" (Chandler 2012).

Research Opportunities

Lynchpin for Development

The willingness of DOE to make part of its federal property available to a commercial energy generation facility is predicated on the research component of the project. As described earlier in the concept of design, there are two major areas of research associated with the Long Island Solar Farm, each of which is outlined in CRADAs.

CRADA 1—The inclusion of data collection instrumentation on the Long Island Solar Farm

Key Research Areas:

- Variability and nondispatchability

- Grid integration of PV solar generation

- Environmental and ecological impacts of large-scale PV solar arrays

CRADA 2—Development of the NSERC

CRADA 1: Data Collection at the Long Island Solar Farm
Overview

The general idea for the CRADA 1 research was to develop and maintain a stream of data from the solar array with a much higher level of granularity than what is tracked throughout the industry. Typically, solar arrays record conditions at intervals of approximately 15 minutes. In contrast, this project was designed to collect data down to the minute or even seconds in some cases. The intent is to compile very detailed data to inform statistical modeling of cloud formations, weather patterns, and general environmental conditions. This will enable researchers to analyze and predict conditional effects on the power performance of the array (Lofaro 2012).

Armed with such vast amounts of data, BNL researchers will be on the cutting edge of climate forecasting research, which could result in new ways to accurately predict power capabilities in the next 30 minutes, the next several hours, or even the next day. This could have tremendously positive impacts: first, it would greatly enhance the solar farm's integration to the local grid, and second, it will help pave the way for further development of solar energy in the East by adding more predictability and certainty to investors.

One of the primary problems with solar energy (and other forms of renewable energy in general) is that sunshine is highly variable and difficult to predict; it just isn't sunny all the time. Obviously at night, the solar farm does not produce any electricity, which is very predictable. The real challenge, though, is the unpredictability of the solar resource at any given time on any given day. PV panels operate very simply—when sunlight hits them, they generate electricity. However, when the sunlight is diffused by clouds, snow, dust, or any other condition, then the output of the array fluctuates precipitously.

Consider a partly cloudy day, for example. As clouds pass over the array—or even just a portion of the array—the power output of the system falls off. As soon as the clouds blow over, the array will ramp back up to full power output instantaneously.

These very steep ramp rates pose a significant challenge to utilities, who cannot predict the output of the solar farm. This makes it impossible for utilities or system operators to rely on the power coming off the array for any meaningful period of time. Figure 13 demonstrates the amount of sunlight received by the LISF in a predictable bell curve over the course of a sunny day. This is a stark contrast Figure 14, which depicts the irradiance measured in the same place on a cloudy day, whereby that nice bell curve collapses unpredictably. Zooming into the 11:00 am peak period also highlights the substantial difference in irradiance even among different sections of the array (each colored line denotes measurements taken by separate pyranometers throughout the solar farm). It is easy to see from these charts how dynamic and complex this variability can be.

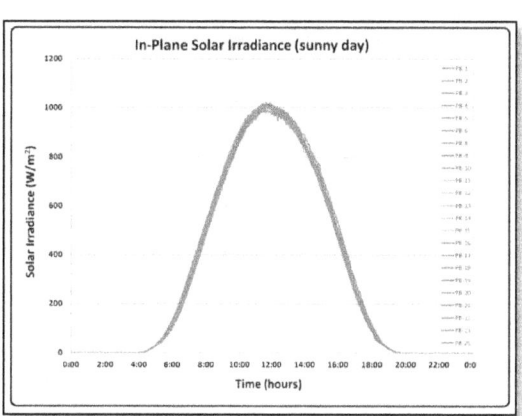

Figure 13. Amount of sunlight received by LISF on a sunny day

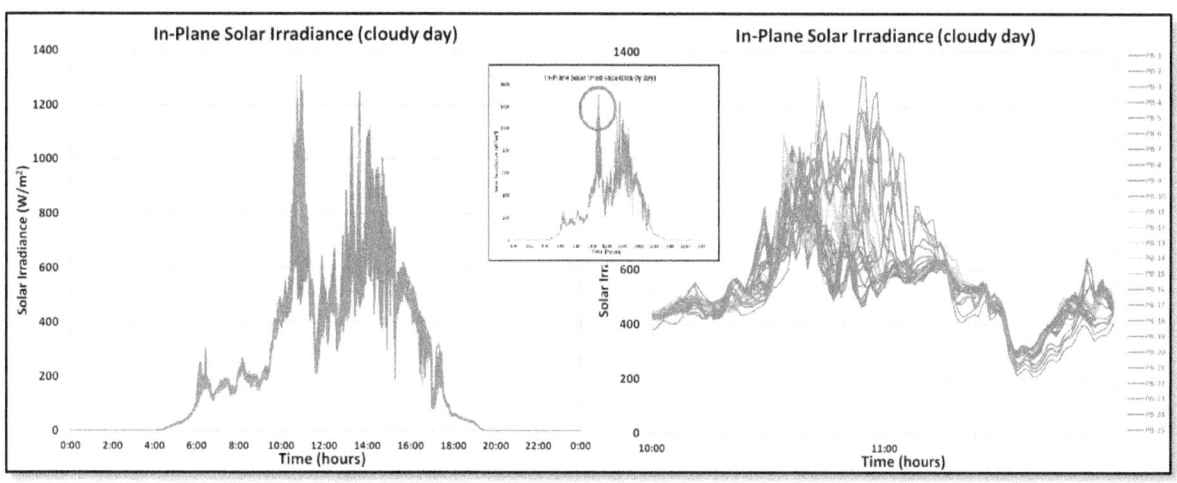

Figure 14. Amount of sunlight measured at LISF on a cloudy day

This variability is a fundamental problem for the PV solar industry. It is important to note, though, that the steep ramp rates and unpredictability of solar power do not have a dramatic impact on the grid. This is because solar energy represents only a minute amount of power compared to the rest of the full energy mix. The LISF, for example, contributes only about 1% of the power supply for Long Island (Deering 2012). This should not diminish the importance of this issue, though. In order for solar energy (and other highly variable resources) to increase its percentage contribution to the overall generation portfolio, its predictability and reliability must be dramatically improved. This poses a cutting-edge problem that the LISF project is well-equipped to help solve.

The major goals of CRADA 1's key areas include the following (BNL Solar Research Agenda 2010):

1) Variability and nondispatchability:

- Develop database with time-resolved solar resource and meteorological data for BNL (Northeast United States) and solar plant electrical performance

- Study the effects of weather variations and cloud transients on large-scale system output

- Develop models for forecasting of available solar irradiation and impact on power production (minutes, hours, day ahead)

- Examine relationship between solar radiation input and solar plant output

- Participate in the NREL PV Variability Database.

2) Grid Integration of PV solar generation:

- Establish the performance characteristics of large-scale solar PV plants located in the Northeast and factors affecting capacity credit

- Evaluate impacts of large-scale solar PV arrays on the reliability and stability of the grid and local distribution systems

- Characterize the response of various system elements (e.g., inverter recovery) following rapid changes in solar insolation due to cloud transients and how it impacts array output

- Develop models to predict the response of the solar plant and individual components (e.g., inverters) to fault conditions

- Evaluate the impact of surface soiling on solar plant performance

- Study long-term performance and reliability of system and components

- Firm LIPA's capacity credit for the LISF.

3) Environmental and ecological impacts of large-scale PV solar arrays:

- Perform life-cycle assessments of large-scale solar PV plants and energy policy options

- Study the effect of the array to the local environment, including microclimate, meteorology, and hydrology, and the associated impacts to the habitats for local plants and animals

- Collect data useful for evaluating the impacts of large-scale solar plants to regional and global climate, including greenhouse gas emissions

- Observe and document impacts to local plant and animal life near the large array.

CRADA 2: The Northeast Solar Energy Research Center (NSERC)

Overview

The second part of the research agenda is the development of the dedicated research array, the NSERC. As shown in Figure 15, this is a completely separate and much smaller array than the solar farm. The NSERC will be used as a testing ground for all the different component parts of solar power technology. This 1 MW array would generate power that would remain behind the meter at Brookhaven National Laboratory, helping the lab move toward its internal sustainability goals (Lofaro 2012). As of this publication, the land has been cleared for development, but the NSERC is still in its design phase. Construction is set to take place in the fall, and the NSERC is expected to be open for research in early 2013.

Figure 15. Location of the Northeast Solar Energy Research Center

Modular Design

The concept of the NSERC makes it a unique proving ground for solar energy research and technology. The entire research center will be separated into three "solar array areas," each of which will be configured for different research functionalities. Generally those are:

- Area 1: Testing inverters and other electronic components (approximately 800 kW)

- Area 2: Testing tracking technologies and structural engineering (approximately 100 kW)

- Area 3: Testing new PV module technologies (approximately 100 kW)

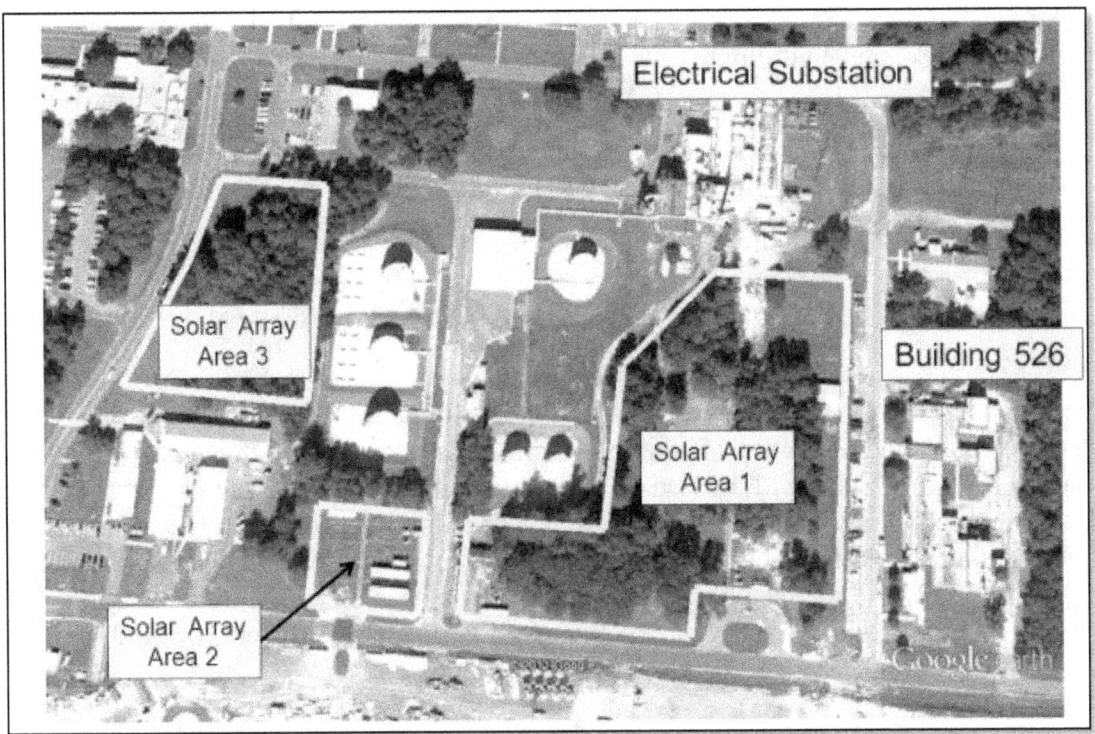

Figure 16. Development Plan for the NSERC Arrays

Each of these arrays will have a modular design to the greatest extent possible. This will enable users to test innovations in various component parts of the system. The "plug and play" concept adds a great deal of technical complexity to the design and naturally increases the overall design costs. What it will provide, though, is an extremely valuable opportunity. It will be a unique

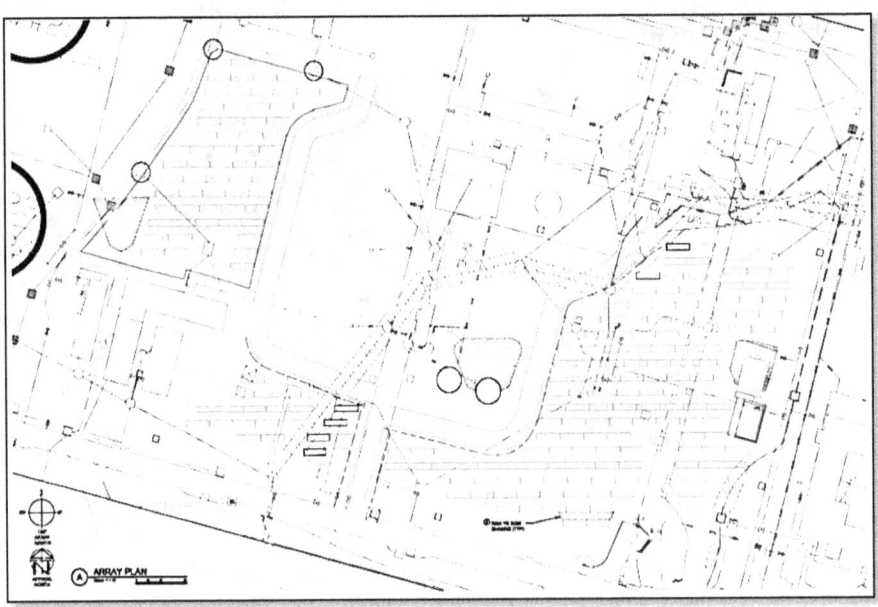

Figure 17. NSERC Array plan

place for industry to come and test new solar technologies. Often times it is prohibitively difficult for companies or individuals to get through the permitting process just to test new ideas. The NSERC could be that perfect incubator test bed for all sorts of new ideas to develop (Chandler 2012). The NSERC will set those conditions for industry. It will enable researchers to prove new technologies and conduct crucial research to improve the efficiency and deployability of PV solar in the Eastern United States, and more broadly.

40

No Specific Agenda

It is important to realize that the research component of the solar farm project is not a research project unto itself, but rather it is the development of a laboratory facility to conduct a myriad of research projects *for* industry, *by* industry. Brookhaven does not have particular research that it wants to conduct for its own purposes. BNL's intention is to establish a resource for industry to conduct cutting edge research and development for solar energy (Lofaro 2012). Below is a sample menu of research areas that NSERC is intended to support:

<u>Solar Energy Research:</u> Standardized testing of solar system components:

- PV panel technology
- Inverters and micro-inverters
- Converters and micro-converters
- Advanced metering infrastructure

<u>Storage Systems Integration</u>

- Batteries
- Mechanical storage
- Chemical storage

<u>Smart Grid Technology Applications and Study</u>

- Modeling
- Advanced metering infrastructure technologies

<u>Solar Simulators and Forecasting Models</u>

- Near term (minutes)
- Midterm (hours)
- Long term (next day/s)

<u>Load Simulators</u>

<u>Failure Analyses and Vulnerability Mitigation</u>

<u>National PV Environmental Research Center</u>

- Environmental, health, and safety aspects of photovoltaic systems
- Life-cycle analyses

<u>Meteorological Services Program</u>

- On-site met services

Results

Though the project is relatively new, and in many ways has only just begun, there are many different ways to measure the results coming out of the Long Island Solar Farm. As of this writing, the NSERC has yet to be built, so it has yet to produce results [other than the fact that it has generated a lot of attention from the PV solar industry and energy research groups (Lofaro, Looney 2012)]. The large array has been up and running since November 2011, and the preliminary results already indicate resounding success. The project as a whole has been remarkably beneficial for each of the parties involved.

The Public

All said and done, this project presents real value for a reasonable price. Michael Hervey, chief operating officer of LIPA, explains that, "[The cost amounts to] about $0.60 per month for the typical residential customer on Long Island for the next 20 years. In addition, (the power purchase agreement with the LISF, LLC) not only allows us to provide clean energy to LIPA's customers, but it also delivers price stability for our customers in an energy market where oil and gas prices remain volatile" (CCFE, 2012). Though PV solar energy remains comparatively expensive and the financial costs are shouldered by Long Island ratepayers, that price is generally considered affordable compared to the perceived benefits received by the community.

Environmental Groups

The project is a success when considering the opportunity costs to the environment from generating an equivalent amount of electricity from a fossil fuel source. Through that lens, the LISF results in the abatement of more than 30,000 metric tons of carbon dioxide emissions per year, as well as substantial amounts of other pollutants such as nitrogen oxide (NOx) and sulfur dioxide. These environmental benefits, combined with a thoughtful and comprehensive stakeholder engagement approach, helped the project earn the Best Photovoltaic Project of Year Award (for 2011) from the New York Solar Energy Industries Association (NYSEIA).

Atmospheric Impacts

- CO_2 Avoided: 30,950 metric tons/year (1.2 million metric tons over 40 years) Compared to conventional electric generating facilities on Long Island
- CO_2 Sequestration Lost: 842 metric tons/year lost due to removal of trees (33,680 metric tons over 40 years)
- Methane Avoided: 80 metric tons over 40 years

Environmental Benefit Contributions

- BNL formally protecting 89 acres (51 + 38) in site master plan
- LIPA formally protecting 45 acres
- LISF contributing $75,000 to LI Native Plant Initiative for environmental restoration
- $2 million of forgone consideration to preserve additional property

Other Pollutants Avoided

- 1,320 metric tons of Nitrogen Oxides (NO_x) avoided over 40 years
- 11,200 metric tons of Ozone Season NO_x avoided over 40 years
- 3,040 metric tons of Sulfur Dioxide avoided over 40 years
- The energy required to produce and construct arrays equals approximately 2 years of output (88,000 MWh equivalent)

Though the project had a substantial impact on the environment on federal property, DOE, BP Solar, and LIPA made a determined effort to engage with environmental groups, namely the Pine Barrens Society, the Long Island Native Plant Initiative, and the Pine Barrens Commission. The result was the natural resource benefits package, which was a robust suite of mitigation efforts that left environmental groups largely satisfied.

U.S. Department of Energy

The project is a success for DOE, who was able to create a win-win-win scenario whereby the local community benefits, the business sees a profit, and the whole project is in keeping with the Department of Energy's mission. It also demonstrated that a project like this can be done where it was traditionally thought it would be impractical or even impossible: on a Superfund site, and in sensitive environmental conditions, and in the Northeast in general. It accomplished all of this in an innovative way and in keeping with DOE's mission. Members of the DOE at Brookhaven were also awarded Secretarial Achievement Awards for their work on this project.

Brookhaven Science Associates, Brookhaven National Laboratory

The results for Brookhaven National Laboratory have been very positive, and are only just beginning. The unprecedented amount of instrumentation that has been built into the solar farm is already producing huge amounts of data. This is providing uniquely valuable data sets for researchers to analyze. The data is driving innovations in solar forecasting models, large-scale grid integration, and environmental effects, among other areas of research.

At the time of this writing, BNL is still working with BP Solar to develop the NSERC. Once that is operational (projected to be in early 2013), BNL will be able to open its doors to industry as the premier solar research and development facility in the east.

Long Island Power Authority (LIPA)

According to Mike Deering, this is a very successful project and the array is producing power very well. The best results are that, with this one project, LIPA was able to:

- Bring in 32 MW of clean electricity to the grid

- Avoid using fossil energy as fuel with its fluctuating costs and environmental hazards

- Significantly reduce peak energy load

- Diversify LIPA's portfolio with renewable resources on Long Island

- Improve economic development; at the time of peak construction, it was the largest construction project on Long Island

- Generate renewable energy credits toward LIPA's renewable portfolio standard.

Mr. Deering also stated that, "Long Islanders are proud of this project. Accolades for it have come from every diverse corner of Long Island society and interest groups—Businesses, Environmentalists, Democrats, Republicans, and everyone in between" (Deering 2012).

BP Solar

In December 2011—the very next month after the LSIF became operational—BP Solar declared it would be going out of business. This was a financial decision for the broader company based on macro-level decisions, which were a long time in the making. The Long Island Solar Farm did not contribute to that decision; it was a profitable project for BP Solar (Chandler 2012).

According to Richard Chandler, "BP Solar really built a platinum project at BNL with a long-term trajectory in mind. It's a great facility, with a strong partner in DOE and LIPA, and it has a strong balance sheet; all the risk has been taken out. Plus, by the numbers, the array is performing well."

He added that the best aspect about this project is that it's not just another power array. It has increasing value over time because of the research piece. There are opportunities to get great information from the research being done on the large array, and the NSERC is an innovation that will surely create more public-private partnerships to move the industry forward.

Implications and Next Steps

While some of the results of this project are evidently off to a successful start, the real benefits of the Long Island Solar Farm and the NSERC have only just begun. This endeavor is replete with new implications for each of the parties involved.

The Public and Environmental Groups

This project has afforded the community and environmental groups several benefits with regard to generating clean energy, avoiding polluting alternatives, and mitigating its environmental impacts. This project sets a precedent in two overarching ways.

First, this project represents a new opportunity for development. With the establishment of a successful commercial project on federal lands, it is likely that other investors and developers will come knocking at DOE's doorstep. Depending on the types of projects that are proposed or authorized, the effects on the environment and the rest of the community will vary.

Second, this project establishes a very high standard of engagement and interaction between project developers, DOE, and the public. The proponents of the LISF project were very meticulous in their engagement with the community right from the start. They established open lines of communication among all the parties. This kept the public involved and informed, which facilitated the project moving relatively smoothly through the design, permitting, and construction processes. This strategy effectively diffused any major opposition from the public or environmental groups, which could have been easily fomented by disenfranchisement from the process.

Any new project will have to genuinely engage and involve the public in the same way. The community and environmental groups on Long Island not only deserve it, they expect it, too.

U.S. Department of Energy

As the administrator of the federal property at Brookhaven National Laboratory, DOE went out of its comfort zone to make this project a reality. DOE assumed significant risk in offering to host a commercial solar farm, exposing the department to potential backlash from the community and other federal agencies who could claim that this project was, at face value, an illicit use of public lands for private profits.

The DOE demonstrated, however, that a project like this can be done and well within the parameters of the department's mission and in the best interests of the lab. They also demonstrated the importance of executing a thorough public outreach program and ensuring the cooperation of the project developer with an up-front investment of $2.3 million from BP Solar. This insulated the DOE from the business risk that BP Solar would go out of business. Nobody expected they would, but they did. Thus, DOE's insistence on receiving that down payment saved the viability of the project for DOE and the lab from BP Solar, who would have otherwise left the project without a research array. This serves as an important template in this kind of public-private partnership.

The LISF project also very much falls in line with the DOE's Asset Revitalization Initiative, or more formally known as the Energy Parks Initiative. The EPI is designed to leverage "federal assets to increase the taxpayers' return on investment" (Gilbertson 2009). In practical terms, "EPI will convert [Office of Environmental Management] liabilities (e.g., contaminated sites, facilities, and materials) into assets to solve critical national energy issues." This includes, but is not limited to, the repurposing of federal lands for the development of renewable energy[9] (Gilbertson 2009).

In fact, according to Bob Gordon, Brookhaven National Laboratory was approach by people in the [Energy Parks Initiative]. "They were wanting to uphold [The Long Island Solar Farm] as an example of their initiative. This project was not, however, developed under that program, and it preceded the inception of the [Energy Parks Initiative] altogether."

Clearly, the LISF is in the general spirit of what the Department of Energy is trying to do in the future—leverage available federal facilities for the development of innovative energy projects. Though it is not directly associated with the Energy Parks Initiative, the Long Island Solar Farm serves as a successful example of what can be achieved.

Brookhaven Science Associates, Brookhaven National Laboratory

According to Dr. Looney, "The Long Island Solar Farm and the research array have changed the face of BNL to a certain extent. It has become a visible symbol of the future of energy technology and the Brookhaven's role at its cutting edge. The project also created a new relationship between BNL and LIPA, as well as new relationships with DOE/EERE program managers."

---

[9] Other EPI uses for federal lands cited include: renewable energy (solar, wind, biomass, geothermal); fossil fuels (clean coal, gas turbines); electricity generation, transmission, distribution; hydrogen generation; emission controls (carbon sequestration, specialty manufacturing); and nuclear (power, fuel cycle, waste management).

BNL has already been approached by EPRI, who wanted to join us in their solar energy research. National Center for Atmospheric Research (NCAR) also came to us for test-bed data as it relates to solar energy. These are two very clear examples of how major organizations in the PV solar industry recognize the research and development opportunities at Brookhaven (Looney).

Long Island Power Authority (LIPA)

There were some key lessons that LIPA took away from this project, particularly with regard to its experience with the request for proposals. The solar RFP was the first of its kind for LIPA and New York, and LIPA soon discovered that the RFP process was very time and labor intensive. Though LIPA was no stranger to RFPs, this one attracted a great deal of attention. As a result, the whole RFP process took quite a lot of time and staff-work to prepare the request, submit it, field questions, and organize and analyze all the responses. That was all before routing recommendations through LIPA's chain of authorities, which was time consuming in itself.

Another drawback of the RFP process was that, in the public's eye, LIPA ostensibly became associated with the "developers" of the project. The nature of the RFP and the review procedures required LIPA to be very much engaged in the process, more so than they would have preferred to be. Such involvement gave outsiders the general impression that LIPA was a principal developer of the project, when that was not the case (Deering). LIPA's primary interest was the power resource, not any one particular style or program of development.

From those two major lessons, LIPA has moved towards using feed-in tariffs (FITs) as the primary vehicle for connecting PV solar projects to the grid. Instead of outlining individualized parameters solar power projects, LIPA has chosen to standardize the policies by which solar projects, big and small, can connect to the grid. This establishes clarity and certainty for developers and helps LIPA avoid being in the weeds of the development process (Deering 2012).

LIPA generally expects to see more distributed PV solar programs develop across Long Island. It is unlikely that Long Island will see another PV solar array built on a scale close to the LISF, but there is still a lot of potential for double-digit megawatt arrays (Deering 2012).

BP Solar and the PV Solar Industry

BP Solar is committed by a $2.3 million down payment to finishing the development of the NSERC before it goes out of business and disengages completely from working with the lab. Their follow-through with this consideration aspect of the easement agreement will not necessarily benefit BP Solar directly, but it certainly will help set conditions to advance the PV solar industry through research and innovations made at the NSERC.

The implication for investors and developers in the broader PV industry are two-fold. First, the Long Island Solar Farm is a new source of truly vast amounts of information and research capabilities that are unavailable elsewhere, certainly not in the Eastern United States. Second, this project represents an example of an effective public-private partnership that is mutually beneficial and profitable as a commercial enterprise. This project opens new possibilities for industry to work with federal and state agencies in places where there may be the presumption that those organizations would be uninterested in this type of PV solar deployment. Where there is a will, there certainly is a way.

Conclusion

*Is the Long Island Solar Farm
a good template
for utility-scale PV solar development
in the Eastern United States?*

Not a Good Template

According to the Oxford dictionary, the definition of a "template" is "something that serves as a pattern from which other similar things can be made." If we apply this definition to the Long Island Solar Farm it certainly presents a compelling case study as an example of success for a variety of different stakeholders. There were, however, some remarkably nuanced conditions that enabled this project to become a reality. It is the uniqueness of those circumstances and conditions that prevent this project from being imitated in many places, if anywhere else at all.

Some of those unique conditions include:

List A: Unique Circumstances

- No transmission was necessary; LIPA had an existing substation immediately adjacent to the proposed site

- The availability of very large tracts of unused land on federal property in close proximity to load

- The classification of the lab property as a "Superfund site" created a voluminous body of thorough environmental data

- The lab's history as a military installation generated a trove of detailed surveys of historical and cultural artifacts on the property

- The mission of the federal facility was such that a commercial project could be made mutually beneficial

There were also a number of conditions that were absent from the LISF site, any one of which could have derailed the LISF project or prevented it from happening altogether. Some of these include:

List B: Anti-Conditions

- Endangered or protected species of plants or animals (although the endangered tiger salamander is widely prevalent on the site, there was space enough for the project to be built around its protected habitat)

- Legal restrictions on land use within the federal enclave

- An irreconcilable opposition group

Clearly, the conditions in List A represent a combination of very fortunate circumstances that greatly enabled the development of the Solar Farm. If any single one of those conditions did not exist, for example, it could have had dramatic impacts on the time it would take to develop the project, the overall costs, or both. Additionally, the "anti-conditions" in List B are circumstances that could easily be found in many other locations and, if present, would likely prevent a project like this from developing.

The aggregate of these conditions makes the experience of the Long Island Solar Farm very difficult, if not impossible, to replicate elsewhere. Therefore, even though the Long Island Solar Farm has proven to be a success story for large-scale PV solar energy, it does not readily "serve as a pattern from which other similar things can be made." For that reason, this project is not a good template for utility-scale solar development in the East.

A Trailblazing Resource

The Long Island Solar Farm does not necessarily have to be replicable to serve as an example. Instead of *being* a good model *of* development, the LISF provides a unique opportunity to *build* good models *for* development.

The research agenda at Brookhaven National Laboratory, including the vast instrumentation on the solar farm and the dedicated NSERC array, will provide essential data and research opportunities. The instrumentation on the solar farm is already generating valuable information that is supplying the development of better solar forecasting models.

The NSERC will also give industry an unparalleled opportunity to push the limits of PV solar technologies and test their applications in the East. The research agenda at BNL will greatly enhance industry's understanding of the conditions they face in building PV solar installations in the region. By answering some of the major questions for industry, the Long Island Solar Farm will thereby reduce the barriers to future development of major solar projects in the east. So, in the end, even though it may be impossible to literally follow in the LISF's footsteps, industry will certainly be able to find its way down a clearer path forward, thanks to the research being done at the lab.

A Good Example: Innovation in Attitude

The Long Island Solar Farm is certainly a remarkable success story in that it has become a large-scale and mutually beneficial asset for a diverse group of public and private stakeholders. What makes the LISF story so exemplary is that, in all likelihood, it really never should have happened in the first place. As Mike Deering described, the group that came together represented "the right organizations, the right people, and the right attitude to see it through." Dr. Looney agreed that the group was "a real coalition of the willing."

Table 2 shows how unlikely this project was from the start. Each of the proponents chose to get involved and commit to finding success. Nobody would have faulted any of these groups had they remained with their most likely course of action, which was to deny the project.

Table 2. Finding Success in Unlikely Positions

	Most Likely Position (No!)	Actual Position (Yes!)	Risk
LIPA	1) Requesting small-scale, distributed solar projects 2) Rejecting a radical proposal from BP Solar	1) Opening the possibility of a utility-scale project with a bold solar RFP 2) Selecting BP Solar's proposal	1) All eggs in one basket 2) A commercial project on federal land would languish in permitting and public opposition
BP Solar	1) The company only develops smaller projects (10-15 MW) 2) Can only tolerate a lease agreement 3) BP Solar is a business; not interested in supporting a research agenda	1) Proposed a 37 MW array 2) Agreed to an easement 3) Using consideration could add value to BP Solar and support the DOE mission	1) Unfamiliar territory for BP business models, especially with variable conditions in the East 2) Loss of long-term land control 3) Financial loss
DOE	Potential conflict of interests; a federal agency cannot support a commercial development project on public lands	DOE can support a commercial venture if the project fits the mission of the Department of Energy	Public backlash in support of the most likely position
BNL	A commercial project will not fully support a research agenda	Allocating use of consideration to develop research agendas; insured with down payment	Embarrassment for the Lab if research agenda disintegrates or never materializes
Public	DOE and BNL are outside of their purview; should not support commercial development at the national lab	If done properly, the project would benefit the community	Setting a precedent for commercial development at BNL or on other public lands
Environmental Groups	This project comes at a high cost of crucial Pine Barren ecosystems; could incite firestorm of public opposition	Mitigation efforts could be effective to compensate for forested lands being cut down	Setting a precedent for commercial development at a high cost of environmental resources

The table demonstrates how the participation of each of these groups was a distinct departure from what was expected of them. If one of these groups had chosen to adhere to their most likely position, the project would not have gotten off the ground.

In addition to mass—and unlikely—participation, a distillation of the Table 2 down to its essential parts reveals some key conditions and decisions that were utterly crucial to development.

49

<u>List C: Key Conditions and Decisions</u>

- DOE's willingness to host a commercial venture on federal property

- LIPA's selection of BP Solar's radical proposal

- BP Solar's willingness to sign an easement instead of a lease

- BP Solar's willingness to use consideration as a means to fulfill DOE mission requirements

- DOE's willingness to legally enable BP Solar to cash-out of consideration obligations (separate from the down payment on the NSERC)

- A joint effort from LIPA, BP Solar, DOE, and BSA on a robust natural resources benefits package

- A vigorous communication and outreach plan with the public

- Respectful inclusion of all interest groups, even (and especially) those in opposition

Each of the parties involved in the development of the LISF recognized the unusual opportunities that the favorable conditions in Lists A and B presented. This is what helped pull each group from their most likely positions. Furthermore, this helped buttress a sense of teamwork among the disparate groups, which enhanced their collective ability to make decisions and set the conditions to ensure successful completion of the project.

On Leadership

Leadership comes in different forms, of course. In many cases, the most efficient type of leadership results from a single decision-maker. In contrast, the experience of the LISF project does not appear to support that concept.

The development process does not distinguish any single point of leadership that was ultimately responsible for driving this project to completion. You cannot say, for instance, that DOE took the lead on this project because in many ways LIPA did, and so did BP Solar. Even the public and community leaders got involved to shape the debates and outcomes. For example, members of each group cited the work of Congressman Bishop, who was actively engaged and served as a mediator in the development debates (Looney, Gordon, and Deering 2012). While that was important and helpful, Congressman Bishop's participation was not necessarily crucial to the project's success. In the end, though, no group established itself as dominant over the other. Accordingly, on the surface, there appears to be a conspicuous absence of any single point of leadership in this endeavor.

While no single leader can be rightly identified, each stakeholder group certainly exhibited a great deal of leadership within their own camps.

Key features of this internal leadership include:

List D: Leadership Attributes

- Having an ambitious vision for the future

- Recognizing opportunities

- Taking bold action while mitigating risks

- Looking for constructive ways to achieve a positive goal versus looking for reasons not to

- A genuine respect for disparate interests

- The ability to compromise and cooperate

What is striking about the development of the Long Island Solar Farm is that each of the stakeholder groups consistently displayed every one of these essential leadership qualities. Consider for a moment if any of these qualities were absent from any of the stakeholder groups—the effects would have been disastrous for the group effort. In the same vein, if any one group tried to establish itself as the single dominant leader in this consortium of interest groups, it is likely that the project would have failed. The interests that brought each party to the table would have taken on a perception that they were there to support a single group's agenda. This would have diluted the attraction of mutual benefits that each group expected from the outcomes, and the project would have likely collapsed on itself before even getting started.

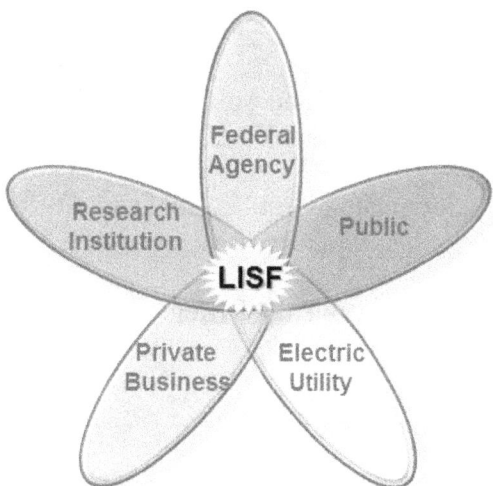

Figure 18. Leadership Star Diagram

In Closing

Undoubtedly, leadership played a quintessential role in the development of the Long Island Solar Farm. Each group indeed had the "right people" who could effectively lead their organizations according to those critical tenets outlined in List D. Those leaders were able to come together to strike a balance of cooperation that structured the debates, safeguarded individual interests, and found a way to make an unlikely dream a reality. That is the essence of the Long Island Solar Farm, and that is what makes it an excellent example for public-private partnerships and the future of solar energy.

51

References

"2011 SEPA Utility Solar Rankings: SEPA's Fifth Annual Report Highlighting Utility Solar Leaders." (2012). *Solar Energy Power Association.* Accessed July 2, 2012: http://www.solarelectricpower.org/solar-tools/sepa-utility-solar-rankings.aspx.

Amper, Richard. (July 13, 2012). Long Island Pine Barrens Society, Riverhead, New York.

Bergin, Tom and Sarah Young (Dec. 21, 2011). "BP Turns Out Lights at Solar Business." *Reuters Online.* Accessed July 10, 2012: http://www.reuters.com/article/2011/12/21/us-bp-solar-idUSTRE7BK1CC20111221.

"BNL Solar Research Agenda: Brookhaven National Laboratory Research Agenda on Solar Photovoltaic Systems." (Sept. 15, 2010). Upton, New York: Brookhaven National Laboratory.

"BP Alternative Energy: Solar Power." (2013). BP Solar. Accessed in 2012: http://www.bp.com/sectiongenericarticle.do?categoryId=9025019&contentId=7046515.

"Brookhaven Science Associates: Laboratory Administration." (2013). Brookhaven National Laboratory. http://www.bnl.gov/bnlweb/admin/bsa.asp.

"Central Pine Barrens Joint Planning and Policy Commission." (2013). Central Pine Barrens Joint Planning and Policy Commission. http://www.pb.state.ny.us/index.htm#Act_Plan_Commission.

Chandler, Richard. (July 16, 2012). Long Island Solar Farm, BP Solar International, Inc., Brookhaven National Laboratory, Upton, New York.

"Citizens Campaign for the Environment: About Us/Contact." (2012). Citizens Campaign for the Environment. http://www.citizenscampaign.org/about.asp.

"Database of State Incentives for Renewables & Efficiency." (2012). DSIRE. Accessed July 13, 2012: http://www.dsireusa.org.

Deering, Michael J. (July 26, 2012). Long Island Power Authority, LIPA Headquarters, Uniondale, New York.

EA-1663. "Environmental Assessment for BP Solar Array Project." (December 2009). Brookhaven Site Office, Office of Science, U.S. Department of Energy.

Easement Outgrant. "For Installation and Operation of a Solar Electric Generating System at Brookhaven National Laboratory." (January 2010). Agreement between the United States Department of Energy and the Long Island Solar Farm, LLC.

Easement MOU. "Easement Memorandum of Understanding for Cooperation in Solar Energy Between BP Solar International, Inc. and the United States Department of Energy." (February 2010). Attachment 15 to the Easement Outgrant between the United States Department of Energy and the Long Island Solar Farm, LLC.

"Environmental Considerations Map." (2010). *Research Agenda on Solar Photovoltaic Systems.* Upton, New York: Brookhaven National Laboratory.

"EPA Region 2 Superfund: Brookhaven National Laboratory." (2011). Environmental Protection Agency. Accessed Aug. 8, 2012: http://www.epa.gov/region2/superfund/npl/brookhaven/.

EPIA. (May 2012). "Global Market Outlook for Photovoltaics Until 2016." *European Photovoltaic Industry Association.* Accessed July 10, 2012: http://files.epia.org/files/Global-Market-Outlook-2016.pdf.

FERC. (May 2011). "Other Markets: Renewables & Energy Efficiency—Generation & Efficiency Standards." *Federal Energy Regulatory Commission Online.* Accessed Aug. 8, 2012: http://www.ferc.gov/market-oversight/othr-mkts/renew.asp.

Geary, P. (May 25, 2005). "Solar Power Profitability: BP Solar." *Environmental News Network.* Accessed July 10, 2012: http://www.climateark.org/shared/reader/welcome.aspx?linkid=42124.

Gilbertson, M. A. (April 23, 2009). "Energy Parks Initiative." The Office of the Deputy Assistant Secretary for Engineering & Technology, U.S. Department of Energy. Accessed July 14, 2012: http://www.em.doe.gov/pdfs/Gilbertson_EnergyParkInitiativeUpdate.pdf.

Gordon, R. P. (June 27, 2012). Brookhaven Site Office, Office of Science, U.S. Department of Energy. Brookhaven National Laboratory, Upton, New York.

Green, T. M. (July 17, 2012). Environmental Protection Division, Brookhaven Science Associates. Brookhaven National Laboratory, Upton, New York.

Lenardic, D. (July 13, 2012). "Large-Scale Photovoltaic Power Plants, Ranking 1-50." *PV Resources.* Accessed July 18, 2012: http://www.pvresources.com/PVPowerPlants/Top50.aspx.

"LIPA, BP Solar and Brookhaven National Lab Flip the Switch at the Long Island Solar Farm." Long Island Power Authority. Accessed July 11, 2012: http://www.lipower.org/newscenter/pr/2011/111811-solar.html.

"LIPA Chooses BP, enXco for Solar Installations." (March 19, 2012). *Sustainable Business.com.* Accessed July 10, 2012: http://www.sustainablebusiness.com/index.cfm/go/news.display/id/17827.

"LIPA: Company Profile." (2013). Long Island Power Authority. http://www.lipower.org/company/profile/.

"LIPA Electric Resource Plan 2010-2020." (February 2010). Long Island Power Authority. Accessed July 10, 2012: http://www.lipower.org/company/powering/energyplan10.html.

"LIPA Energy Plan 2004-2013, Volume 2. Energy Primer." (June 23, 2004). Long Island Power Authority. http://www.lipower.org/pdfs/company/projects/v1.energyplan.pdf.

Lofaro, R. J. (July 10, 2012). Renewable Energy Group, Sustainable Energy Technologies Department, Brookhaven Science Associates. Brookhaven National Laboratory, Upton, New York.

"Long Island Solar Farm Largest in State History." (Nov. 21, 2011). Citizens Campaign for the Environment. Accessed July 13, 2012: http://www.citizenscampaign.org/news/story.asp?id=465.

Looney, P.J. (July 17, 2012). Sustainable Energy Technologies Department, Brookhaven Science Associates. Brookhaven National Laboratory, Upton, New York.

"New York State Department of Environmental Conservation: Eastern Tiger Salamander Fact Sheet." Accessed Aug. 6, 2012: http://www.dec.ny.gov/animals/7143.html.

"NREL: Dynamic Maps, GIS Data, & Analysis Tools." National Renewable Energy Laboratory, Golden, CO. Accessed July 10, 2012: http://www.nrel.gov/gis/solar.html.

Ryan, M. (Aug. 31, 2011). "Tax Break Tactics Divide Renewable Industry." *AOL Energy.* Accessed July 10, 2012: http://energy.aol.com/2011/08/31/tax-break-tactics-divide-renewable-industry/.

Sadler, L. (July 19, 2012). Brookhaven Site Office, Office of Science, U.S. Department of Energy. Brookhaven National Laboratory, Upton, New York.

"SERDP/ESTCP: Energy and Water." Strategic Environmental Research and Development Program (SERDP) and Environmental Security Technology Certification Program (ESTCP). Accessed July 5, 2012: http://www.serdp-estcp.org/Program-Areas/Energy-and-Water.

"SMUD's PV1: Solar Power for 25 Years." (Sept. 9, 2009). *Green Energy News.* Accessed July 19, 2012: http://www.green-energy-news.com/nwslnks/clips909/sep09011.html.

Snook, J. (March 7, 2007). "Department of Defense Completes Renewable Energy Assessment." U.S. Department of Energy—Energy Efficiency and Renewable Energy, Federal Energy Management Program. Accessed July 12, 2012: http://www1.eere.energy.gov/femp/news/printable_versions/news_detail.html?news_id=10612

"The Long Island Pine Barrens Society." Long Island Pine Barren Society. http://pinebarrens.org.

"United States Department of Energy: Office of Science." (2013). U.S. Department of Energy. http://science.energy.gov/laboratories/site-offices/.

Wiegand, P. (2011). "Long Island Native Plant Initiative." Friends of the Bay. Retrieved on July 21, 2012 from: http://friendsofthebay.org/?p=1200.

Wood, E. (June 20, 2012). "Solar in New York: Its Strategy to Make Solar Shine" *Renewable Energy World.Com.* Accessed July 19, 2012: http://www.renewableenergyworld.com/rea/news/article/2012/06/solar-in-new-york-its-strategy-to-make-solar-shine.